T0134920

Lecture Notes in Electrical Engineering

Volume 981

The book series *Lecture Notes in Electrical Engineering* (LNEE) publishes the latest developments in Electrical Engineering—quickly, informally and in high quality. While original research reported in proceedings and monographs has traditionally formed the core of LNEE, we also encourage authors to submit books devoted to supporting student education and professional training in the various fields and applications areas of electrical engineering. The series cover classical and emerging topics concerning:

- Communication Engineering, Information Theory and Networks
- Electronics Engineering and Microelectronics
- Signal, Image and Speech Processing
- Wireless and Mobile Communication
- Circuits and Systems
- Energy Systems, Power Electronics and Electrical Machines
- Electro-optical Engineering
- Instrumentation Engineering
- Avionics Engineering
- Control Systems
- Internet-of-Things and Cybersecurity
- Biomedical Devices, MEMS and NEMS

For general information about this book series, comments or suggestions, please contact leontina.dicecco@springer.com.

To submit a proposal or request further information, please contact the Publishing Editor in your country:

China

Jasmine Dou, Editor (jasmine.dou@springer.com)

India, Japan, Rest of Asia

Swati Meherishi, Editorial Director (Swati.Meherishi@springer.com)

Southeast Asia, Australia, New Zealand

Ramesh Nath Premnath, Editor (ramesh.premnath@springernature.com)

USA, Canada

Michael Luby, Senior Editor (michael.luby@springer.com)

All other Countries

Leontina Di Cecco, Senior Editor (leontina.dicecco@springer.com)

** This series is indexed by EI Compendex and Scopus databases. **

Humberto Jesús Corona Pampín · Reza Shirvany
Editors

Recommender Systems in Fashion and Retail

Proceedings of the Fourth Workshop at the Recommender Systems Conference (2022)

Springer

Editors
Humberto Jesús Corona Pampín
Spotify
Amsterdam, The Netherlands

Reza Shirvany
Digital Experience-AI and Builder Platform
Zalando SE
Berlin, Germany

ISSN 1876-1100   ISSN 1876-1119 (electronic)
Lecture Notes in Electrical Engineering
ISBN 978-3-031-22194-1   ISBN 978-3-031-22192-7 (eBook)
https://doi.org/10.1007/978-3-031-22192-7

This Springer imprint is published by the registered company Springer Nature Switzerland AG
The registered company address is: Gewerbestrasse 11, 6330 Cham, Switzerland

Preface

Online fashion retailers have significantly increased in popularity over the last decade, making it possible for customers to explore hundreds of thousands of products without the need to visit multiple stores or stand in long queues for checkout. However, the customers still face several hurdles with current online shopping solutions. For example, customers often feel overwhelmed with the large selection of the assortment and brands. In addition, there is still a lack of effective suggestions capable of satisfying customers' style preferences, or size and fit needs, necessary to enable them in their decision-making process. In this context, recommender systems are very well positioned to play a crucial role in creating a highly adaptive and personalized customer experience in fashion. Moreover, in recent years, social shopping in fashion has surfaced, thanks to platforms such as Instagram, providing a very interesting opportunity that allows to explore fashion in radically new ways. Such recent developments provide exciting challenges for the recommender systems and machine learning research communities:

(1) In fashion e-commerce, the impact of social networks and the influence that fashion influencers have on the choices people make for shopping is undeniable. For instance, many people use Instagram to learn about fashion trends from top influencers which helps them to buy similar or even exact outfits from the tagged brands in the posts. This approach can extensively facilitate customers quest for outfits. When traced, customers' social behaviour can be a very useful guide for online shopping Websites, providing insights on the styles the customers are really interested in, and hence aiding the online shops in offering better recommendations. However, a challenge not yet tackled is how to effectively leverage such implicit signals. For example, some customers might like certain outfits found in social media for inspiration, but factors such as budget, body shape, and other implicit or explicit preferences might need to be taken into account in order to make relevant recommendations for the customers.

(2) Another well-known difficulty with recommendation of similar items is the large quantities of clothing items which can be similar, but belong to different brands. Such items may also have very different price points or sustainability

characteristics. Relying only on implicit customer behavioural data—such as the user's purchase history, viewed or favorited items—will not be sufficient in the coming future to distinguish between a recommendation that will lead to an item being purchased and kept, vs. a recommendation that might result in either the customer not following it, or eventually return the item. In these cases, advanced content-aware solutions need to be developed. For example, with the advent of deep learning and computer vision techniques, outfits and product images and their associated descriptions can be analysed to understand the interesting style details that attract the customers. The sourcing of metadata derived from images is a well-known challenge which often leads to the manual tagging of fashion items at the content production stage, leading to poor-quality medatada, and thus reducing the quality of the recommendation and search systems.

(3) Finding online clothes that fit is also very challenging. In fact, finding the right size and fit is one of the major factors not only impacting customers purchase decision, but also their satisfaction from e-commerce fashion platforms. In this context, customers need to purchase garment and shoes without trying them on. Thus, the sensory feedback phase about how the article fits via touch and visual cues is delayed leading to uncertainties in the buying process and hurdle in returning articles that customers love but don't fit. Therefore, many of consumers remain reluctant to engage in the purchase process in particular for new articles and brands they are not familiar with. To make matters worse, fashion articles have important sizing variations due to for example: coarse definition of size systems (e.g small, medium, large for garments); different specifications for the same size according to the brand; vanity sizing where brands adapt their nominal sizes to target a specific segment of customers based on age, sportiness, etc.; and different ways of converting a local size system to another. Moreover, customer preferences towards perceived article size and fit for their body remain highly personal and subjective which influences the definition of the right size for each customer. The combination of the above factors leaves the customers alone to face a highly challenging problem of determining the right size and fit during their purchase journey, which in turn has resulted in having more than one third of apparel returns to be caused by not ordering the right article size. This challenge presents a huge opportunity for research in intelligent size and fit recommendation systems and machine learning solutions with direct impact on both customer satisfaction and business profitability.

(4) During the past two years, online fashion and retail have been largely influenced by pandemic era events and radical changes in customer and business needs. Some of such effects may only remain on a short term (e.g. restriction measures and lack of physical access to offline shops); however, long-lasting effects in the expectation and customer behaviour have already emerged in the fashion domain whether be it on the choice of the assortment or potential seasonal shifts in online fashion versus offline fashion or even virtual fashion. Although such pandemic-related effects are still rather observed than predicted and not yet fully understood, this challenge presents a radially new opportunity for research in

intelligent recommendation systems and machine learning solutions that can cope with radical changes and/or consider long-term effects of such changes in their predictions.

The *Fourth Workshop on Recommender Systems in Fashion and Retail—fashionXrecsys2022* brought together researchers and practitioners in the fashion, recommendations and machine learning domains to discuss open problems in the aforementioned areas: topics for which the scientific community still needs substantial collaboration between academia and industry, and where new problems show potential but are not broad enough to be discussed in the main conference with enough level of detail. This involves addressing interdisciplinary problems with all of the challenges it entails. Within this workshop, we aimed to start the conversation among professionals in the fashion and e-commerce industries and recommender systems scientists, and create a new space for collaboration between these communities necessary for tackling these deep problems. To provide rich opportunities to share opinions and experience in such an emerging field, we accepted paper submissions on established and novel ideas, as well as new interactive participation formats.

Seattle, USA   Humberto Jesús Corona Pampín
October 2022   Reza Shirvany

Contents

Identification of Fine-Grained Fit Information from Customer Reviews in Fashion

Yevgeniy Puzikov, Sonia Pecenakova, Rodrigo Weffer, Leonidas Lefakis, and Reza Shirvany

Abstract Size recommendation is a task of predicting the best-fitting fashion article size for a customer, based on their purchase history and other signals. *Size* recommendations assume that for every article there is a size that fits the customer's body. However, *fit* recommendations require more fine-grained information to be able to predict the fitting experience and to steer customers to appropriate articles. This requires an understanding of the shape and expected fit of an article, which are not always readily available. In this study, we show how customer reviews can be leveraged to identify topics corresponding to customer concerns and to common issues of the item purchase and post-purchase experience. We empirically evaluate three typologically different topic extraction methods (Latent Dirichlet Allocation, text embedding clustering and zero-shot text classification) in the highly complex domain of real-world fashion reviews and analyse their pros and cons. We demonstrate that coarse-grained topic labels correspond to item's fit issues, such as "too big" and "too small" and show how fine-grained topics signal other specific fit issues of an article (e.g. "short leg length" for jeans or "loose sleeve area" for upper garments).

Keywords Size and fit · Fashion e-commerce · Natural language processing · Topic modelling · Zero-shot text classification

Y. Puzikov (✉) · S. Pecenakova · R. Weffer · L. Lefakis · R. Shirvany
Zalando SE, Tamara-Danz-Str. 1, 10243 Berlin, Germany
e-mail: yevgeniy.puzikov@zalando.de

S. Pecenakova
e-mail: sonia.pecenakova@zalando.de

R. Weffer
e-mail: rodrigo.weffer@zalando.de

L. Lefakis
e-mail: leonidas.lefakis@zalando.de

R. Shirvany
e-mail: reza.shirvany@zalando.de

© The Author(s), under exclusive license to Springer Nature Switzerland AG 2023 1
H. J. Corona Pampín and R. Shirvany (eds.), *Recommender Systems in Fashion and Retail*,
Lecture Notes in Electrical Engineering 981,
https://doi.org/10.1007/978-3-031-22192-7_1

1   Introduction

When shopping online, customers buy articles without trying them on, which increases the chance of getting a garment that does not fit perfectly. In such cases, customers usually return the purchased articles, sometimes losing the incentive to engage in the purchasing process. Offering accurate recommendations in the fashion industry leads to better customer experience, as well as higher business profits due to a reduction in the associated returns. At the same time, logistic improvements reduce carbon footprint of fashion e-commerce platforms, thus making them more sustainable and environmentally friendly.

In order to provide size advice, modern e-commerce platforms use various data sources. Broadly speaking, they can be divided into product-specific information and customer-related data. The former includes packshot images, brand, manufacturer, material composition, confection size information, etc. This data is usually not publicly available, and it is mainly used by industrial researchers to build in-house recommendation systems. Customer-related information includes a variety of data: biometric and billing information, purchase history, customer reviews, etc. Purchase histories constitute the most common data source experimented with in academic studies of recommendation systems. Surprisingly, customer reviews in the fashion domain have been largely ignored: with the exception of some very recent works [3, 10], review texts have been confined to the task of text classification [18, 28, 59, 70], text generation [36, 69] and style transfer [27, 37, 61], with most work focusing on restaurant or movie domains. In this work, we show that customer reviews can be used to extract valuable information and motivate the usage of the extracted knowledge to improve fashion recommendation systems.

Using a novel real-world dataset, we explore how we can understand more about customer's fit preferences that would allow size advice to move from confection size-based recommendations to a more sophisticated fit-aware recommendations system. Fit recommendations require more fine-grained information to predict the fitting experience, compared to the complex but coarser task of predicting the best size. In this study, we show how customer reviews can be leveraged to identify the topics corresponding to customer concerns and to common issues of the item purchase, and post-purchase, experience. Following an in-depth review of the state-of-the-art (SOTA) in the field, we identified three promising, and typologically different, methods: Latent Dirichlet Allocation (LDA), text embedding clustering and zero-shot text classification. We empirically test the utility of these methods, showing how the methods cope with the task of topic extraction and topic prediction on various levels of granularity.

Our contributions: We approach the task of size advice from an unconventional angle by intersecting the article-based fit-related information with the customer view on fit in the form of article reviews, thus exploring the utility of customer reviews in identifying fit-related topics. We compare the performance of typologically different approaches, highlighting their strengths, weaknesses and caveats, like evaluation issues in the case of embedding clustering and the sensitivity of zero-shot text classification approaches to the prompt definition in the fashion domain. Finally, we show how one can perform topic extraction/prediction on different granularity levels. We do that both across article categories and when focusing on a specific category of jeans, a challenging commodity group with complex fit characteristics.

2 Related Work

In this section, we mention most relevant work on developing recommendation systems in fashion industry, as well as the Natural Language Processing (NLP) techniques that can be used to extract topic-like information from customer review texts.

2.1 Recommendation in Fashion

Many recent works in the online fashion domain have focused on the challenging problem of recommending the right size to a customer.

2.1.1 Article-Based Size Advice

Leveraging product-specific information, article-based size advice helps customers make informed decisions by providing information about the article size behaviour, e.g. "this article runs small". Nestler et al. [42] propose a Bayesian model using aggregated article return data together with human expert feedback to predict whether an article is likely to be too small, too large or true to size. Some of the recent studies focus on detecting potential size issues and leverage article images using a deep learning computer vision model [30, 31]. With the aim of unifying the different sizes across size systems and brands, Du et al. [15] propose a method to automate size normalization to a common latent space through the use of order data, generating mapping of any article size to a common space in which sizes can be better compared.

2.1.2 Personalized Size Advice

Another line of research focused on methods to provide a personalized size advice based on customer-related data. Sembium et al. [52] proposed a latent factor model, later followed-up by a Bayesian approach [53], to deduce size features that are classified into ordinal fit predictions (small, fit, or large). Guigourès et al. [20] introduced a hierarchical Bayesian approach which jointly models a size purchased by a customer together with the outcome (kept item or size-related return). A deep learning approach that embeds articles and customers in a latent space to provide a personalized size recommendation has been introduced in various works [14, 54]. Recently,

Lasserre et al. [33] proposed a meta-learning deep learning approach to allow absorbing new data points efficiently without the need for re-training. Hajjar et al. [21] used an attention-based deep learning model, casting the size recommendation problem as a "translation" from articles to sizes.

2.1.3   Fit Advice

Most of the introduced works focus solely on recommending the customer the best size. However, the article's fit characteristics might play an important role in how the customers perceive it. Most existing methods that go beyond size recommendation and explore the domain of fit are based on computer vision algorithms: Hsiao and Grauman [25] introduced body-aware embeddings based on article and fashion model images to identify garments that will flatter specific body types, whereas Pecenakova et al. [44] proposed a generative model to encode and disentangle fit properties of fashion articles. Other approaches explore providing high-fidelity visualization of an article's fit on a body using 3D technologies [4, 26, 43].

2.1.4   Using Customer Reviews

Customer textual reviews, which are offered on most online fashion platforms, can offer an in-depth insight into how an article's fit is perceived by a customer, as shown in [3], where the authors used NLP techniques to extract product fit feedback from customer reviews. Recently, Chatterjee et al. [10] proposed an approach which can use such extracted information from customer reviews as an additional feature along with customer and product features. However, it is used solely to improve size recommendations, rather than understanding specific fit needs of the customers.

2.2   Topic Extraction Techniques

Extraction of useful information from customer reviews can be framed in several ways. One can focus on identifying topic-level relations between words, assuming that a single document covers a small set of concise topics (topic modelling). Another way is to embed each review into a high-dimensional latent space and cluster the resultant text embeddings; one would then manually examine the clusters and assign topic labels. One could also identify potential pain points related to fit they are broadly interested in—e.g. sleeve length, upper/lower garment disproportion—and use those as labels in a zero- or few-shot text classification setting. Below we provide an overview of the SOTA approaches to our task from these three angles.

2.2.1 Unsupervised Probabilistic Models

Since customer reviews do not have any topic labels, it is natural to resort to unsupervised methods of topic modelling. In this regard, the most popular methods include LDA, Dirichlet Mixture Model (DMM), Latent Semantic Analysis (LSA) and dimensionality reduction techniques. LDA [6] is a generative probabilistic model algorithm which assumes that documents are represented as random mixtures over latent topics, where each topic is characterized by a distribution over all the words in the document collection. DMM [67] is similar to LDA, but is aimed at detecting topics in smaller documents and assumes that each text is sampled from only one latent topic. LSA [32] learns topics by first forming a term-document matrix (bag-of-words document representation) and then smoothing the frequency counts to enhance the weight of informative words. This smoothed term-by-document matrix is further decomposed in order to generalize observed relations between words and documents. Dimensionality reduction techniques can be used to project sparse high-dimensional document representations to a lower-dimensional space and induce latent topic assignments; in the past, both Principled Component Analysis (PCA) and Random Projections (RP) have been used for topic modelling [12, 23, 29]. Which of these methods is preferable, depends on many factors: for example, availability of pre-trained embedding models for the target language, or input data properties, like text length. Stevens et al. [57] found that LDA learns the most descriptive topics, while LSA is best at creating a compact semantic representation of documents and words in a corpus. Albalawi et al. [1] found that LDA and LDA with non-negative matrix factorization deliver more meaningful extracted topics and obtain better results than other competing approaches. The same study suggests that the choice of the method depends on the length of the textual data: when one needs to identify topics in a single document or in a small collection of documents, the authors suggest using word frequency-based methods. On the other hand, Yan et al. [65] points out that short texts are too short to capture document-level word co-occurrence statistics, which leads to poor clustering performance of frequency-based approaches to topic modelling.

Given the evidence found in the research literature, we focus on LDA amongst the word frequency-based topic extraction methods: it is scalable, fast, stable in terms of performance and produces meaningful results in most studies.

2.2.2 Clustering Text Embeddings

Word frequency statistics that are used by the conventional topic extraction approaches mentioned above do not consider the position and context of a word in a sentence. Word embeddings, pre-trained on large textual data, offer an alternative that is able to capture contextual information. Early attempts to discover topics for short texts focused on leveraging semantic information encoded in word embeddings after pre-training them on large existing textual data sources, such as GoogleNews and Wikipedia [35, 47, 64]. Currently the SOTA approaches to creating text representations are based on variants of transformer-based models [62]: in particular, Bidirectional Encoder Representations from Transformers (BERT) [11] and its derivatives. Pre-trained BERT models produce high-quality text representations that incorporate the position and context of a word in a sentence. Subakti et al. [58] compared BERT representations to term frequency-inverse document frequency (TF-IDF), a frequency-based statistic that is often used as a sparse text representation, across several clustering algorithms and found that the former outperforms the latter in 28 out of 36 metrics. Fine-tuning BERT representations for downstream tasks is very common in NLP. Unsurprisingly, Yin et al. [66] found this approach to improve the results of short text clustering by 14%.

The general clustering pipeline we experiment with in this work is:

1. Use a pre-trained BERT model to encode the review texts.
2. Apply dimensionality reduction technique (e.g. Uniform Manifold Approximation and Projection (UMAP) [40]).
3. Perform density-based clustering of the embedding vectors (in this work, we use the HDBSCAN [8] clustering algorithm).

2.2.3   Zero-Shot Text Classification

As mentioned in Sect. 1, knowledge extraction from customer reviews could be done via zero-shot text classification. Zero-shot learning is the task of training a classifier on one set of labels and then evaluating it on a different set of labels that the classifier has never seen before [63]. A common approach is to use an existing feature extractor to embed an input and a label name into their corresponding latent representations [56]. In the text domain, one has the advantage that a single model can embed both the input text and the label name into the same latent space, eliminating the need for any alignment step. Pooled word vectors can be used as latent space representations; in the modern NLP context they usually come from pre-trained language models, e.g. BERT or Generative Pre-trained Transformer (GPT) [7, 46].

A popular zero-shot approach was proposed by Yin et al. [68] who used a pre-trained BERT-based Natural Language Inference (NLI) classifier as an out-of-the-box zero-shot text classifier. NLI task considers two sentences: a "premise" and a "hypothesis". The task is to determine whether the hypothesis is true (entailment) or false (contradiction) given the premise (Table 1). The idea proposed by Yin et al.

Table 1 Natural Language Inference: examples of entailing, contradicting and neutral text pairs

Premise	Hypothesis	Label
Physical activity is good for your health	Doing sports makes you healthier	Entailment
	Doing sports makes you a better person	Neutral
	Doing sports is unhealthy	Contradiction

[68] is to take the text we are interested in labelling as the "premise" and to turn each candidate label into a "hypothesis". If the model predicts that the premise "entails" the hypothesis, we take the label to be true. A big advantage of this approach is that it requires minimal effort to apply for the task of topic extraction. More generally, the hypothesis (or modified label) is called a "prompt", and a model is queried to predict a match between the input text and the prompt.

A more general formulation of this setup is exemplified by Task-Aware Representation of Sentences for Generic Text Classification (TARS) proposed by Halder et al. [22]. The authors used a pre-trained model, like BERT and added a binary classification layer on top. They proposed to define M task-specific classification labels and append each of them as part of the input to the model. One then fine-tunes the model to predict "1" for the relations that hold, and "0" for those that do not. For example, for the sentiment classification task, the setup would look as follows:

$$< \text{positive sentiment} > \quad \text{I enjoyed the movie a lot.} \; \rightarrow \; \text{"1"}$$
$$< \text{negative sentiment} > \quad \text{I enjoyed the movie a lot.} \; \rightarrow \; \text{"0"}$$

In this work, we experiment with both TARS and NLI approaches used to directly classify if a review text mentions some fit-related issue of an article.

3 Experiments

We first compare LDA and the embedding clustering pipeline: both are unsupervised methods of information extraction, and their performance can be evaluated using the same metrics. Zero-shot text classification experiments are described in the second half of this section.

3.1 LDA Versus Text Embedding Clustering

3.1.1 Data

In this experiment, we use two sources of data: an anonymized dataset of article reviews written by customers for a variety of fashion articles from a large e-commerce platform and the descriptions of the intended fit characteristics of these articles.

Customer review dataset The dataset consists of 15M reviews in various languages for articles from 39 commodity groups (CG: "jeans", "dress", "shirt", etc.). The non-English reviews were automatically translated into English with the help of a publicly available translation model [60] and an open-source translation library.[1]

---

[1] https://github.com/UKPLab/EasyNMT.

We focus on a subset of 10116 reviews that have been written in the English language originally. Unless otherwise stated, we use this smaller English subset in our experiments.

The reviews are short (2–50 tokens long), have either a neutral or negative rating (≤ 3, on a scale from 1 to 5).[2] Sample reviews are given below:

- "Fits big and the sleeves are way to puffy" (sic)
- "Lovely quality, really oversized"
- "Leggings too long, but quality quite nice"
- "Not a good fit and cheap fabric."

We lower-cased the reviews and cleaned them from emojis and URLs. We do not perform any additional input pre-processing for the embedding clustering approach, as the approaches based on pre-trained BERT-style models operate on raw texts, and any additional pre-processing steps may harm the results [2]. However, LDA is known to require more elaborate data preparation [9, 24], which is why before training LDA models, we additionally lemmatized the reviews and kept only the lemmas which are alphanumeric strings longer than one character, are not in the stopword list and are nouns, adjectives, verbs or adverbs.[3]

Fit and shape annotations Understanding how an article's intended fit relates to the customer's perception as both, a single article, and as part of an outfit is an essential part of moving towards fit-based size advice. In order to explore this connection, we have gathered descriptions of the intended fit characteristics of the articles discussed in customer reviews. The article is described using two terms, "fit" and "shape", and the dataset is produced by a pool of human annotators, that are regularly trained by experts with domain knowledge. The fit of an article describes the article's intended distance from the body and can be ordered into five levels (skinny, slim, regular, loose and oversized), from closest to body to furthest. The shape of an article describes the silhouette or the cut and can be one of six defined categories (fitted, straight, flared, tapered, cocoon, body-hugging). An example of each of these attributes can be seen in Fig. 1. Some descriptive statistics of the used data are shown in Fig. 2.

3.1.2 Evaluation Metrics

As unsupervised methods, LDA and text clustering embedding are evaluated using *topic coherence*. It measures the degree of semantic similarity between the high-scoring words in the discovered topic: the higher the value of the metric, the more

---

[2] We deliberately removed reviews with a rating ≥ 3, because we found that they rarely contain fit-related customer complaints, which is what we want to be able to detect.

[3] We used the English stopword list from Natural Language Processing Toolkit (NLTK) [5]. part-of-speech (POS) tags were obtained with the help of spaCy, a Python library for linguistic analysis of textual data (https://spacy.io/).

Fig. 1 Examples of fit and shape labels. Example images for the different fit levels of jeans are shown in the top row, shape categories are given in the bottom row

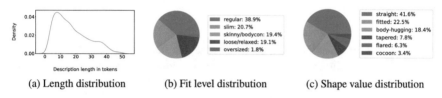

(a) Length distribution   (b) Fit level distribution   (c) Shape value distribution

Fig. 2 Descriptive statistics of the data used in our experiments: distribution of the review lengths, fit levels and shape values

coherent the induced topic is. There are several definitions of the confirmation measures proposed in the literature; in our experiments, we use the so-called C_v measure which was found to be the best formulation according to a systematic study of the configuration space of coherence measures by Röder [49]. Measuring topic coherence is possible if a trained model can generate a bag-of-words (BOW) representation of a topic by extracting the top-N most representative topic words. Text clustering methods by default do not have this capability, since they operate in dense embedding spaces. We use the *class-based TF-IDF* procedure proposed by Grootendorst [19] to model the importance of words in clusters instead of individual documents. This allows one to generate topic-word distributions for each cluster of documents.

3.1.3 Results

Figure 3 shows the average topic coherence scores achieved by LDA and the embedding clustering approach, across a varying number of induced topics. We tested three variants of the text embedding clustering pipeline, depending on which embeddings the model used: vanilla pre-trained model (EMB), same model but fine-tuned on 10K English customer reviews (EMB_ft10K) and the model fine-tuned on 15M multilingual customer reviews auto-translated to English (EMB_ft15M).

Fig. 3 Mean C_v coherence measure scores for LDA and the text embedding clustering pipeline. The score variance across five runs is shown as a band around each plot line. Reference corpus for topic coherence computation: original review texts

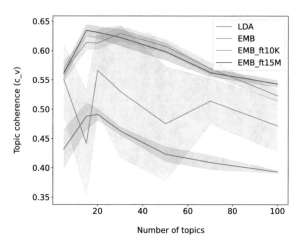

The scores for the embedding pipeline are higher than those for LDA, even in the case of the embedding model which was not fine-tuned on the in-distribution data (EMB). The better performance of the embedding clustering pipeline is consistent with previous studies which showed the superiority of rich embedding representations, compared to sparse BOW representations [58, 66]. Fine-tuning on a small dataset leads to better scores overall, but also a very large performance variance. This finding is consistent with the recent studies of [13, 34, 41] who suggested several explanations of this phenomenon; one of the proposed solutions is increasing the size of fine-tuning data. We tested this hypothesis and found that, indeed, expanding the fine-tuning dataset greatly reduces the variance. However, score-wise, there is barely any performance gain compared to the pre-trained model without any fine-tuning.

3.1.4 Analysis and Discussion

There are several important observations and conclusions which we would like to share.

Identified topics Manual examination of the outputs shows that both approaches detect around seven broad topic categories (wearing experience, commodity group, delivery issues, specific fit area, fabric complaints, comments on seasonality, size-related issues) which have their own more fine-grained components. For example, the category "delivery issues" is further fragmented into topics "received wrong item", "not delivered", "received damaged item".

One advantage of HDBSCAN—the clustering algorithm used by the embedding clustering pipeline—is that it finds the optimal number of clusters on its own. In our experiments, it varied in the range of 34–38. The algorithm also identifies the points which do not belong to any cluster and can be considered as outliers. Figure 4a shows 2D projections of the review embeddings, coloured according to the cluster

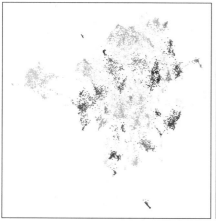
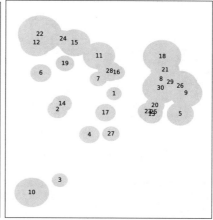

(a) Customer review embedding clusters   (b) Topic clusters identified by LDA

Fig. 4 Visualization of the extracted topics: LDA topics and UMAP-projected review text embeddings clustered by HDBSCAN

Fig. 5 Distribution of the number of topics discussed per individual review in the customer reviews dataset

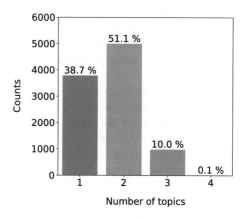

assignments. For comparison, Fig. 4b also shows a visualization of a trained LDA model with 30 induced topics.[4]

Number of topics discussed in individual reviews LDA represents (a) documents as mixtures of topics and (b) topics as mixtures of words. We processed the data using a trained LDA model and counted the number of salient topics assigned to each review. We chose to consider a topic as salient if its proportion in the topic mixture is above 0.2. Figure 5 shows that more than 60% of reviews are multi-topic texts. LDA captures such topic mixtures naturally, because this assumption is baked into the approach.

---

[4] For visualization, we used a Python port of LDAvis [55]: https://github.com/bmabey/pyLDAvis.

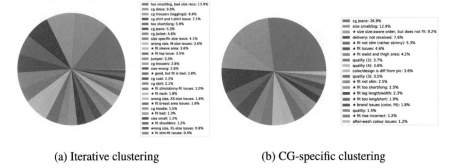

(a) Iterative clustering   (b) CG-specific clustering

Fig. 6 Topic distribution after iterative clustering and CG-specific clustering. The number of topics in both cases was determined automatically by the clustering algorithm. Topics marked with a ★ are fit-related

Discovering fit-related topics Empirically, we observed that increasing the number of topics leads to a more fine-grained, but CG-related clustering (reviews about upper garments would be fragmented into texts about jackets, jumpers, shirts, etc.). However, for the purpose of this work we are more interested in discovering review groups with fit/shape-related issues. To achieve this, we designed the following two strategies: *iterative* and *CG-specific* topic extraction. In what follows, we describe our findings when using the embedding clustering pipeline; LDA models produced similar results.

The iterative topic extraction strategy is as follows: after extracting topics on full data and obtaining coarse-grained topic assignments, we manually examined the text clusters and extracted the reviews which were assigned to fit/shape-related topics, then re-clustered the extracted reviews. Figure 6a visualizes the distribution of topics identified by the embedding clustering pipeline. As can be seen, the strategy indeed allowed us to identify a few topics which were missing in the first experiment. For example, we were able to identify specific fit-critical areas (sleeves, neck, breast, shoulders).

The motivation for the CG-specific topic extraction strategy comes from the fact that fit-related issues can vary largely for different commodity groups. For example, the right fit at the waist might be very important for jeans, but much less of a focus for T-shirts. The idea behind the second strategy is to only cluster reviews that are related to a specific commodity group. We chose to focus on jeans, because it is a challenging clothing category with complex fit characteristics. Figure 6b visualizes the distribution of topics identified by the text embedding clustering pipeline, when using this strategy. It narrows down the topics and extracts valuable fit-related topics, like fit issues in the waist and thigh areas, or leg length being too short or long. However, because many reviews are topic mixtures and because customers are often not very specific in their reviews, the model still induced topics which are prominent, but are unrelated to fit (e.g. material quality or delivery issues).

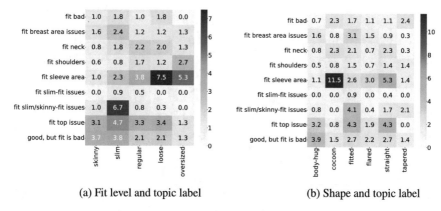

(a) Fit level and topic label   (b) Shape and topic label

Fig. 7 Joint distribution of fit levels/shape values and topic labels predicted by an embedding clustering pipeline using the iterative clustering strategy. We show only fit-related topics. The numbers in the figure are percentages of the total number of customer reviews

Fit-topic joint distribution While extracting topics from reviews can provide information about how the article's fit is perceived by a certain customer, we would further like to study how the article's intended fit relates to the customer's fit experience. For this, we identify which fit and shape attributes co-occur with the various topic assignments. Figure 7 shows the most common fit/shape-related topics common for each of the fit levels and shape values.

We see that loose and oversized fit articles more often have sleeve length issues. A common group of complaints arise when customers purchase slim-fit jeans, but receive an article which in their opinion has skinny fit. Upper-body garments have a higher chance of customer dissatisfaction, if they have slim fit. Oversized items have a higher chance of having problems in the shoulder area. Complimentary observations can be extracted from the shape-topic correlations: for example, cocoon articles often have fit issues in the sleeve area and body-hugging articles sometimes have issues in the upper garment area. These are all very valuable insights which can potentially drive further improvement in existing size recommendation algorithms or even provide an opportunity for new, fit-driven, fashion recommendations. Combined with other article information about brand, manufacturers or material composition, one could determine further potential connections between customer perception and article characteristics, for example, if fit issues correlate with the elasticity of a fabric; or identify brands and manufacturers which produce articles with unusual fit characteristics.

3.1.5 On Evaluation Using Topic Coherence

We would like to highlight several observations regarding evaluation of topic extraction approaches using topic coherence.

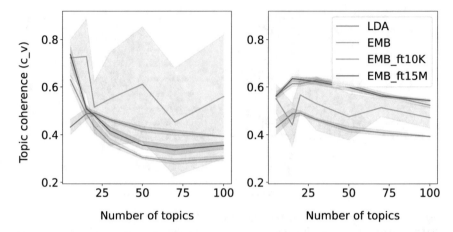

Fig. 8 Mean C_v coherence measure scores for LDA and the text embedding clustering pipeline. Note how the performance of the embedding clustering pipeline changes, depending on what we use as a reference corpus for metric computation: a collection of document clusters, or "super-documents" (**left**), or the original documents (**right**)

Topic coherence and number of topics It is common for clustering methods to achieve better coherence scores when increasing the number of clusters, as this naturally improves the fit of the data. However, we have not seen such improvement when increasing the topic number in our clustering experiments. We explain this behaviour by the fact that most reviews discuss more than one topic, which makes it hard to separate reviews into coherent groups.

Topic coherence and reference corpus definition We found that the results of evaluation using topic coherence greatly depend on which reference corpus is used for topic coherence computation. In our initial experiments, we used an implementation which generated misleadingly poor topic coherence scores (left plot in Fig. 8). Empirically, we traced this issue back to the way the reference corpus is built during the topic coherence score computation. For example, it is possible to use a collection of "super-documents", where each super-document is a concatenation of all documents (reviews, in our case) which were clustered together. Another option is to use a collection of the input documents when computing the topic coherence scores (right plot in Fig. 8). The latter produces higher scores and generally more stable results.

Topic coherence and BOW induction method There are other issues with topic coherence to be aware of. For example, when applied to embedding clustering methods, topic coherence evaluates not only the quality of the topic label assignment, but also the way an approach in question induces BOW representations of topics. A recent study of [71] found a considerable metric score difference (up to 15 C_v score points) between various ways of inducing BOW topic representations. One should be aware of this fact, when evaluating the results of the embedding clustering pipeline or similar approaches.

3.1.6 Summary

Qualitative results suggest that both of the compared methods can be used to discover meaningful topics in customer reviews, but quantitative comparison with LDA shows that the embedding clustering pipeline produces more accurate results, as measured by the topic coherence metric. Finding fine-grained fit-related topics proved to be challenging due to the fact that most customer reviews are mixtures of topics, but the proposed iterative and CG-specific strategies mitigate this problem to some extent and allow one to detect most prominent fit issues. We also assessed the utility of the extracted topics downstream, showing how their intersection with fit and shape article annotations produces insights into the customer experience of fit characteristics of the corresponding garments.

3.2 Zero-Shot Text Classification

So far, we have observed that LDA is a good baseline for extracting fit-relevant information from customer reviews: as shown, most review texts discuss multiple topics and LDA captures such topic mixtures naturally. Even better results can be obtained by using a text embedding clustering pipeline which automatically finds the optimal number of topics and produces more coherent topic clusters.

However, there are two big disadvantages of both approaches. First, they require a manual examination of the clusters to assign a meaningful label. This makes automating and scaling the topic extraction pipeline rather difficult and error-prone. Second, out of all clusters, only a few relate to fit, and the rest are of limited use, if one wants to focus on extracting fit-related topics.

Both of these problems can be potentially addressed by employing large pre-trained language models in a zero-shot text classification scenario: one can define the labels of interest and use the models to predict if an issue denoted by the label is salient for a given review. This section describes our experiments with two recent zero-shot methods: TARS and NLI classifier.

3.2.1 Data

In order to evaluate a zero-shot text classification system, one needs an annotated dataset. To make the experiment more constrained, we decided to restrict ourselves to the commodity group of jeans, due to its high fit complexity. Furthermore, we focus on three specific garment areas which often surface in the customer reviews as problematic (hips, waist and rise) and annotate them from the perspective of having a "too big" or "too small" issue. We collected 120 data points in total, 20 for each of the (area, issue) combination, annotated by two human experts in the domain of fit.

3.2.2   Evaluation Metrics

Since the zero-shot systems we experiment with are text classifiers, we use standard text classification metrics for evaluation: precision (P), recall (R) and their harmonic mean, F1-score (F1). We do not use topic coherence metric in this experiment, as it does not provide any valuable information regarding the methods' downstream performance. The text classifiers we evaluate use BERT-based text embedding modules which are similar to the one employed by the text embedding clustering pipeline, so the topic coherence scores would be similar to the ones obtained by the text embedding clustering pipeline.

3.2.3   Results

We use three classification setups with a varying level of label granularity: predicting if the article has a size issue (two labels : "too big" and "too small"); if a specific fit-critical area has an issue (three labels : "hips", "waist" and "rise"); and a combination of the two, resulting in a total of six class labels. Table 2 shows the results of metric evaluation of the two zero-shot approaches tested. A random baseline is added for comparison.

The NLI approach clearly outperforms both TARS and the random baseline. For a large number of inputs, TARS failed to output any prediction: it did not find any class label to be matching the input review text. To avoid breaking the evaluation pipeline, for such cases we implemented a fallback option of predicting a match uniformly at random among the labels in the label set. It turned out, TARS very often followed the fallback option: 83.1% of predictions in the case of the binary classification setup, 56.5% when predicting if an article area has a fit issue and 79.8% in the 6-class setting.

Table 2 Evaluation results of the zero-shot text classification approaches applied to the task of predicting fit-related issues in customer reviews

Label set	Model	Precision	Recall	F1-Score	Accuracy
Size issue (2)	TARS	0.62	0.55	0.47	0.54
	NLI	0.67	0.67	**0.67**	0.67
	Random	0.53	0.53	0.53	0.53
Area issue (3)	TARS	0.25	0.39	0.30	0.39
	NLI	0.40	0.36	**0.35**	0.37
	Random	0.32	0.32	0.32	0.31
Area and size issue (6)	TARS	0.13	0.18	0.09	0.18
	NLI	0.34	0.34	**0.33**	0.34
	Random	0.19	0.18	0.18	0.18

We use three classification setups with a varying number of classes (2, 3 and 6). Precision, recall and F1-score are macro-averaged (all classes are treated equally)

Table 3 Prompt definition sensitivity of zero-shot classifiers

Prompt	Modification rationale	Output probability
too big	No modification (original label)	0.94
these jeans are too big	Short phrase	0.34
These jeans are too big	Upper-cased phrase	0.38
These jeans are too big.	Sentence	0.25
it is too big	A phrase closer to the input	0.80
It's too wide on the bottom.	Same as input	0.99
On bottom. too the it's wide	Shuffled input	0.99
big	Short label	0.80
too	Short non-sensical (but input-overlapping) label	0.92
It's not too wide on the bottom.	Add negation	0.00

In this example, a zero-shot NLI text classifier outputs a probability of a prompt entailing the input sentence "It's too wide on the bottom." The original prompt is "too big". The modified versions, the rationale behind the change and the corresponding probability of a match are given in the first, second and third columns, respectively

One potential reason as to why both zero-shot approaches did not largely outperform the baseline could be the domain shift between the training data used to train the embedding models and our customer review dataset. Evaluation of the benefits of fine-tuning for the zero-shot approaches is reserved for future work.

Label definition sensitivity Zero-shot models are known to be sensitive to the prompt definition [16, 38, 39, 45, 48, 50]. We tested this hypothesis on our real-world dataset by randomly selecting a few reviews and one simple label ("too big"), creating small modifications of the label by paraphrasing and changing word order and then passing the modified label together with the review text to the NLI classifier which performed best in our experiments. Ultimately, we wanted to observe whether the predicted probability of a given label changes (Table 3). For illustrative purposes, we showcase our experiment on one specific review ("It's too wide on the bottom."), but similar results were obtained when inspecting other review texts.

The above example illustrates the challenges of predicting fit-related issues using a zero-shot text classifier. Even a slight change of the prompt, like adding a full-stop or a word which also happens to be in the input, drastically changes the prediction. One can scramble the words in the sentence and still score high.

3.2.4 Summary

In our study, out-of-the-box zero-shot text classification models exhibited large prediction variance: depending on the prompt definition, the model returns vastly different results. There exist techniques to mitigate this—for example, Schick and Schötze [51] proposed to use an ensemble of several models using several prompts, while

Gong and Eldardiry [17] design prompts based on an external knowledge graph to integrate semantic knowledge information. Fine-tuning the zero-shot model before applying to the review data could potentially improve the results, as the language of the reviews and the prompted labels are different from the data the pre-trained BERT-based models were trained on. We leave experimentation with these and other approaches for future work.

Utility of extracted topics How can extracted topics solicit building better recommendation systems? Customer-specific experience of fit recommendations is not in scope of this work; we focus more on detecting article problems using aggregated customer feedback in the form of topic labels. Therefore, direct application of the extracted topics lies in the area of article fit advice.

First of all, the topics can be incorporated as prior information for the recommendation systems which rely on the aggregated article return data to predict whether an article has size issues (e.g. SizeFlags [42]): extracted topics would help to stabilize the flags by providing more evidence to the algorithm. One could also use the topics as an additional feature for systems which detect garment areas with potential fit and shape issues (as is done by SizeNet [30]): customer reviews marked as having fit issues often specify the area which does not fit right. Finally, meta-learning approaches, like MetalSF [33], could benefit from embedding the predicted topic labels into the latent space shared with the size labels. One can use either the embedding of a single topic assignment, or a weighted average of the embeddings corresponding to the predicted topic distribution, as an additional input to the system. Alternatively, one could also use a label vector representing the counts of the various size-related clusters which the reviews for a specific article belong to. In both cases, this would result in enriching the article-related information available to the model, leading to more accurate recommendations.

4   Conclusion

In this work, we take the first step towards combining article-based fit-related data with the customer perception of fit in the form of textual reviews. The empirical results suggest that the topics extracted from the customer review texts, combined with the article fit annotations, can be leveraged to identify fit issues and further incorporated as additional inputs into size advice systems.

Our evaluation of three typologically different methods for topic extraction from fashion article customer reviews suggest that, compared to LDA, a text embedding clustering pipeline produces more accurate results, as measured by the topic coherence metric. To further narrow down the several distinct, broad topic categories extracted by both approaches, an iterative clustering and a commodity group-specific clustering strategy was developed. Both strategies allow a model to identify fit-related topics with finer granularity. However, two challenges make the task difficult: reviews often touch upon several topics and the customers are often not very specific in their reviews, which can result in extracted topics which are salient and distinct, but not related to fit issues.

In addition, the empirical evidence showed that the performance of fine-tuned large pre-trained language models depends on the size of the data used for fine-tuning: smaller fine-tuning datasets produce models with larger performance variance. Several caveats of using topic coherence for topic extraction evaluation were identified: the choice of the reference corpus and of the method to induce BOW topic representations needs to be carefully considered in order to produce solid results. Insights were provided on how article-specific fit and shape attributes are related to certain fit-related topics that customers experience and how this can help identify commodity groups which are prone to specific fit issues. Further insights can be obtained by intersecting the extracted topics with other product-related information, like brand and manufacturer or material composition. Compared to LDA and the embedding clustering approaches, our study suggests zero-shot text classification to be a less mature technology. While promising, such classifiers are too sensitive to the prompt definition, which makes them very unreliable in practice with (at times messy) real-world data.

Evaluation of the various methods automating prompt definition, as well as the few-shot text classification scenario, are reserved for future work. Looking forward, we also plan to evaluate the benefits of multilingual embedding models, which can help reduce any potential inaccuracies of the translation process, that the current monolingual embedding pipeline requires.

Acknowledgements We thank Matti Lyra, Adrien Renahy and Josip Krapac for their contribution on the data topics and the related discussions, Ximena Hernandez Garibaldi and the Fashion experts for greatly assisting us in the human expert annotations, Hareesh Pallikara Bahuleyan and Hamid Zafar for their fruitful feedback on our work.

References

1. Albalawi R, Yeap TH, Benyoucef M (2020) Using topic modeling methods for short-text data: a comparative analysis. Front Artif Intell 3
2. Alzahrani E, Jololian L (2021) How different text-preprocessing techniques using the BERT model affect the gender profiling of authors. CoRR abs/2109.13890, https://arxiv.org/abs/2109.13890
3. Baier S (2019) Analyzing customer feedback for product fit prediction. arXiv preprint arXiv:1908.10896
4. Bhatnagar BL, Tiwari G, Theobalt C, Pons-Moll G (2019) Multi-garment net: learning to dress 3D people from figures. In: Proceedings of the IEEE/CVF international conference on computer vision, pp 5420–5430
5. Bird S, Klein E, Loper E (2009) Natural language processing with Python: analyzing text with the natural language toolkit. O'Reilly, Beijing. http://www.nltk.org/book
6. Blei DM, Ng AY, Jordan MI (2003) Latent Dirichlet allocation. J Mach Learn Res 3(null):993–1022
7. Brown TB, Mann B, Ryder N, Subbiah M, Kaplan J, Dhariwal P, Neelakantan A, Shyam P, Sastry G, Askell A, Agarwal S, Herbert-Voss A, Krueger G, Henighan T, Child R, Ramesh A, Ziegler DM, Wu J, Winter C, Hesse C, Chen M, Sigler E, Litwin M, Gray S, Chess B, Clark J, Berner C, McCandlish S, Radford A, Sutskever I, Amodei D (2020) Language models are few-shot learners. CoRR abs/2005.14165, https://arxiv.org/abs/2005.14165

8. Campello RJGB, Moulavi D, Sander J (2013) Density-based clustering based on hierarchical density estimates. In: Pei J, Tseng VS, Cao L, Motoda H, Xu G (eds) Advances in knowledge discovery and data mining. Springer, Berlin Heidelberg, Berlin, Heidelberg, pp 160–172
9. Celard P, Vieira AS, Iglesias EL, Borrajo L (2020) LDA filter: a latent Dirichlet allocation preprocess method for Weka. PLOS ONE 15(11):1–14
10. Chatterjee O, Tej JR, Dasaraju NV (2022) Incorporating customer reviews in size and fit recommendation systems for fashion e-commerce. In: Proceedings of the the 2022 SIGIR workshop on ecommerce. Association for Computing Machinery
11. Devlin J, Chang MW, Lee K, Toutanova K (2018) Bert: pre-training of deep bidirectional transformers for language understanding. arXiv preprint arXiv:1810.04805
12. Ding W, Hossein Rohban M, Ishwar P, Saligrama V (2013) Topic discovery through data dependent and random projections. In: Dasgupta S, McAllester D (eds) Proceedings of the 30th international conference on machine learning, PMLR, Atlanta, Georgia, USA, Proceedings of machine learning research, vol 28, pp 1202–1210. https://proceedings.mlr.press/v28/ding13.html
13. Dodge J, Ilharco G, Schwartz R, Farhadi A, Hajishirzi H, Smith NA (2020) Fine-tuning pretrained language models: weight initializations, data orders, and early stopping. CoRR abs/2002.06305, https://arxiv.org/abs/2002.06305
14. Dogani K, Tomassetti M, De Cnudde S, Vargas S, Chamberlain B (2019) Learning embeddings for product size recommendations. In: SIGIR eCom, Paris, France. https://sigir-ecom.github.io/ecom19Papers/paper13.pdf
15. Du ES, Liu C, Wayne DH (2019) Automated fashion size normalization. ArXiv abs/1908.09980
16. Gao T, Fisch A, Chen D (2020) Making pre-trained language models better few-shot learners. CoRR abs/2012.15723. https://arxiv.org/abs/2012.15723
17. Gong J, Eldardiry H (2021) Prompt-based zero-shot relation classification with semantic knowledge augmentation. CoRR abs/2112.04539. https://arxiv.org/abs/2112.04539
18. Gräbner D, Zanker M, Fliedl G, Fuchs M (2012) Classification of customer reviews based on sentiment analysis. In: Fuchs M, Ricci F, Cantoni L (eds) Information and communication technologies in tourism 2012. Springer, Vienna, pp 460–470
19. Grootendorst M (2022) BERTopic: neural topic modeling with a class-based TF-IDF procedure. https://doi.org/10.48550/ARXIV.2203.05794. https://arxiv.org/abs/2203.05794
20. Guigourès R, Ho YK, Koriagin E, Sheikh AS, Bergmann U, Shirvany R (2018) A hierarchical Bayesian model for size recommendation in fashion. In: Proceedings of the 12th ACM conference on recommender systems. ACM, pp 392–396
21. Hajjar K, Lasserre J, Zhao A, Shirvany R (2020) Attention gets you the right size and fit in fashion. In: Submitted to the 14th ACM conference on recommender systems. ACM
22. Halder K, Akbik A, Krapac J, Vollgraf R (2020) Task-aware representation of sentences for generic text classification. In: Proceedings of the 28th international conference on computational linguistics. International Committee on Computational Linguistics, Barcelona, Spain (Online), pp 3202–3213. https://doi.org/10.18653/v1/2020.coling-main.285. https://aclanthology.org/2020.coling-main.285
23. Hecking T, Leydesdorff L (2018) Topic modelling of empirical text corpora: validity, reliability, and reproducibility in comparison to semantic maps. CoRR abs/1806.01045, http://arxiv.org/abs/1806.01045
24. Hoyle AM, Goel P, Resnik P (2020) Improving neural topic models using knowledge distillation. In: Proceedings of the 2020 conference on empirical methods in natural language processing (EMNLP). Association for Computational Linguistics, Online, pp 1752–1771. https://doi.org/10.18653/v1/2020.emnlp-main.137. https://aclanthology.org/2020.emnlp-main.137
25. Hsiao WL, Grauman K (2020) Vibe: dressing for diverse body shapes. In: Proceedings of the IEEE/CVF conference on computer vision and pattern recognition (CVPR 2020), pp 11059–11069
26. Januszkiewicz M, Parker C, Hayes S, Gill S (2017) Online virtual fit is not yet fit for purpose: an analysis of fashion e-commerce interfaces. In: Proceedings of the 8th international conference and exhibition on 3D body scanning and processing technologies, pp 210–217. https://doi.org/10.15221/17.210

27. Jin D, Jin Z, Hu Z, Vechtomova O, Mihalcea R (2022) Deep learning for text style transfer: a survey. Comput Linguist 48(1):155–205
28. Johnson R, Zhang T (2017) Deep pyramid convolutional neural networks for text categorization. In: Proceedings of the 55th annual meeting of the Association for Computational Linguistics (Volume 1: Long Papers). Association for Computational Linguistics, Vancouver, Canada, pp 562–570. https://doi.org/10.18653/v1/P17-1052. https://aclanthology.org/P17-1052
29. Kanerva P, Kristofersson J, Holst A (2000) Random indexing of text samples for latent semantic analysis. In: Gleitman L, Josh A (eds) Proceedings of the 22nd annual conference of the Cognitive Science Society, Erlbaum, New Jersey, vol 1036
30. Karessli N, Guigourès R, Shirvany R (2019) Sizenet: weakly supervised learning of visual size and fit in fashion images. In: IEEE conference on computer vision and pattern recognition (CVPR) workshop on FFSS-USAD
31. Karessli N, Guigourès R, Shirvany R (2020) Learning size and fit from fashion images. In: Springer' special issue on Fashion recommender systems
32. Landauer TK, Dumais ST (1997) A solution to Plato's problem: the latent semantic analysis theory of the acquisition, induction, and representation of knowledge. Psychol Rev 104:211–240. http://www.welchco.com/02/14/01/60/96/02/2901.HTM
33. Lasserre J, Sheikh AS, Koriagin E, Bergmann U, Vollgraf R, Shirvany R (2020) Meta-learning for size and fit recommendation in fashion. In: SIAM international conference on data mining (SDM20)
34. Lee C, Cho K, Kang W (2019) Mixout: effective regularization to finetune large-scale pretrained language models. CoRR abs/1909.11299, http://arxiv.org/abs/1909.11299
35. Li C, Wang H, Zhang Z, Sun A, Ma Z (2016) Topic modeling for short texts with auxiliary word embeddings. In: Proceedings of the 39th international ACM SIGIR conference on research and development in information retrieval. Association for Computing Machinery, New York, NY, USA, SIGIR '16, pp 165–174. https://doi.org/10.1145/2911451.2911499
36. Li J, Zhao WX, Wen JR, Song Y (2019) Generating long and informative reviews with aspect-aware coarse-to-fine decoding. In: Proceedings of the 57th annual meeting of the Association for Computational Linguistics, Association for Computational Linguistics, Florence, Italy, pp 1969–1979. https://doi.org/10.18653/v1/P19-1190. https://www.aclweb.org/anthology/P19-1190
37. Li P, Tuzhilin A (2019) Towards controllable and personalized review generation. In: Proceedings of the 2019 conference on empirical methods in natural language processing and the 9th international joint conference on natural language processing (EMNLP-IJCNLP). Association for Computational Linguistics, Hong Kong, China, pp 3237–3245. https://doi.org/10.18653/v1/D19-1319. https://aclanthology.org/D19-1319
38. Liu P, Yuan W, Fu J, Jiang Z, Hayashi H, Neubig G (2021) Pre-train, prompt, and predict: A systematic survey of prompting methods in natural language processing. CoRR abs/2107.13586, https://arxiv.org/abs/2107.13586
39. Lu Y, Bartolo M, Moore A, Riedel S, Stenetorp P (2021) Fantastically ordered prompts and where to find them: overcoming few-shot prompt order sensitivity. CoRR abs/2104.08786, https://arxiv.org/abs/2104.08786
40. McInnes L, Healy J, Melville J (2018) Umap: uniform manifold approximation and projection for dimension reduction. https://doi.org/10.48550/ARXIV.1802.03426. https://arxiv.org/abs/1802.03426
41. Mosbach M, Andriushchenko M, Klakow D (2020) On the stability of fine-tuning BERT: misconceptions, explanations, and strong baselines. CoRR abs/2006.04884, https://arxiv.org/abs/2006.04884
42. Nestler A, Karessli N, Hajjar K, Weffer R, Shirvany R (2020) Sizeflags: reducing size-related returns in fashion e-commerce. In: Submitted to the 14th ACM conference on recommender systems. ACM
43. Patel C, Liao Z, Pons-Moll G (2020) Tailornet: predicting clothing in 3D as a function of human pose, shape and garment style. In: Proceedings of the IEEE/CVF conference on computer vision and pattern recognition, pp 7365–7375

44. Pecenakova S, Karessli N, Shirvany R (2022) Fitgan: fit-and shape-realistic generative adversarial networks for fashion. arXiv preprint arXiv:2206.11768
45. Petroni F, Lewis PSH, Piktus A, Rocktäschel T, Wu Y, Miller AH, Riedel S (2020) How context affects language models' factual predictions. CoRR abs/2005.04611, https://arxiv.org/abs/2005.04611
46. Radford A, Wu J, Child R, Luan D, Amodei D, Sutskever I (2018) Language models are unsupervised multitask learners. CoRR https://d4mucfpksywv.cloudfront.net/better-language-models/language-models.pdf
47. Rangarajan Sridhar VK (2015) Unsupervised topic modeling for short texts using distributed representations of words. In: Proceedings of the 1st workshop on vector space modeling for natural language processing. Association for Computational Linguistics, Denver, Colorado, pp 192–200. https://doi.org/10.3115/v1/W15-1526. https://aclanthology.org/W15-1526
48. Reynolds L, McDonell K (2021) Prompt programming for large language models: beyond the few-shot paradigm. CoRR abs/2102.07350, https://arxiv.org/abs/2102.07350
49. Röder M, Both A, Hinneburg A (2015) Exploring the space of topic coherence measures. In: Proceedings of the eighth ACM international conference on web search and data mining. Association for Computing Machinery, New York, NY, USA, WSDM '15, pp 399–408. https://doi.org/10.1145/2684822.2685324
50. Rubin O, Herzig J, Berant J (2021) Learning to retrieve prompts for in-context learning. CoRR abs/2112.08633, https://arxiv.org/abs/2112.08633
51. Schick T, Schütze H (2021) Exploiting cloze-questions for few-shot text classification and natural language inference. In: Proceedings of the 16th conference of the European chapter of the Association for Computational Linguistics: main volume. Association for Computational Linguistics, Online, pp 255–269. https://doi.org/10.18653/v1/2021.eacl-main.20. https://aclanthology.org/2021.eacl-main.20
52. Sembium V, Rastogi R, Saroop A, Merugu S (2017) Recommending product sizes to customers. In: Proceedings of the eleventh ACM conference on recommender systems. ACM, pp 243–250
53. Sembium V, Rastogi R, Tekumalla L, Saroop A (2018) Bayesian models for product size recommendations. In: Proceedings of the 2018 world wide web conference, WWW '18, pp 679–687
54. Sheikh AS, Guigourès R, Koriagin E, Ho YK, Shirvany R, Bergmann U (2019) A deep learning system for predicting size and fit in fashion e-commerce. In: Proceedings of the 13th ACM conference on recommender systems. ACM
55. Sievert C, Shirley K (2014) LDAvis: a method for visualizing and interpreting topics. In: Proceedings of the workshop on interactive Language learning, visualization, and interfaces. Association for Computational Linguistics, Baltimore, Maryland, USA, pp 63–70. https://doi.org/10.3115/v1/W14-3110. https://aclanthology.org/W14-3110
56. Socher R, Ganjoo M, Manning CD, Ng A (2013) Zero-shot learning through cross-modal transfer. In: Burges C, Bottou L, Welling M, Ghahramani Z, Weinberger K (eds) Advances in neural information processing systems, vol 26. Curran Associates, Inc. https://proceedings.neurips.cc/paper/2013/file/2d6cc4b2d139a53512fb8cbb3086ae2e-Paper.pdf
57. Stevens K, Kegelmeyer P, Andrzejewski D, Buttler D (2012) Exploring topic coherence over many models and many topics. In: Proceedings of the 2012 joint conference on empirical methods in natural language processing and computational natural language learning. Association for Computational Linguistics, Jeju Island, Korea, pp 952–961. https://aclanthology.org/D12-1087
58. Subakti A, Murfi H, Hariadi N (2022) The performance of BERT as data representation of text clustering. J Big Data 9(1):15
59. Sun C, Qiu X, Xu Y, Huang X (2019) How to fine-tune BERT for text classification? In: Sun M, Huang X, Ji H, Liu Z, Liu Y (eds) Chinese computational linguistics. Springer International Publishing, Cham, pp 194–206
60. Tiedemann J, Thottingal S (2020) OPUS-MT—building open translation services for the world. In: Proceedings of the 22nd annual conference of the European Association for Machine Translation (EAMT), Lisbon, Portugal

61. Toshevska M, Gievska S (2021) A review of text style transfer using deep learning. CoRR abs/2109.15144, https://arxiv.org/abs/2109.15144
62. Vaswani A, Shazeer N, Parmar N, Uszkoreit J, Jones L, Gomez AN, Kaiser L, Polosukhin I (2017) Attention is all you need. CoRR abs/1706.03762, http://arxiv.org/abs/1706.03762
63. Wang W, Zheng VW, Yu H, Miao C (2019) A survey of zero-shot learning. ACM Trans Intell Syst Technol (TIST) 10:1–37
64. Xun G, Gopalakrishnan V, Ma F, Li Y, Gao J, Zhang A (2016) Topic discovery for short texts using word embeddings. In: 2016 IEEE 16th international conference on data mining (ICDM), pp 1299–1304. https://doi.org/10.1109/ICDM.2016.0176
65. Yan X, Guo J, Liu S, Cheng X, Wang Y (2013) Learning topics in short texts by non-negative matrix factorization on term correlation matrix, SIAM, pp 749–757. https://doi.org/10.1137/1.9781611972832.83. https://epubs.siam.org/doi/abs/10.1137/1.9781611972832.83
66. Yin H, Song X, Yang S, Huang G, Li J (2021) Representation learning for short text clustering. CoRR abs/2109.09894, https://arxiv.org/abs/2109.09894
67. Yin J, Wang J (2014) A Dirichlet multinomial mixture model-based approach for short text clustering. In: Proceedings of the 20th ACM SIGKDD international conference on knowledge discovery and data mining. Association for Computing Machinery, New York, NY, USA, KDD '14, pp 233–242. https://doi.org/10.1145/2623330.2623715
68. Yin W, Hay J, Roth D (2019) Benchmarking zero-shot text classification: datasets, evaluation and entailment approach. CoRR abs/1909.00161, http://arxiv.org/abs/1909.00161
69. Zang H, Wan X (2017) Towards automatic generation of product reviews from aspect-sentiment scores. In: Proceedings of the 10th international conference on natural language generation. Association for Computational Linguistics, Santiago de Compostela, Spain, pp 168–177. https://doi.org/10.18653/v1/W17-3526. https://aclanthology.org/W17-3526
70. Zhang X, Zhao J, LeCun Y (2015) Character-level convolutional networks for text classification. In: Proceedings of the 28th international conference on neural information processing systems—volume 1. MIT Press, Cambridge, MA, USA, NIPS'15, pp 649–657
71. Zhang Z, Fang M, Chen L, Namazi Rad MR (2022) Is neural topic modelling better than clustering? An empirical study on clustering with contextual embeddings for topics. In: Proceedings of the 2022 conference of the North American chapter of the Association for Computational Linguistics: human language technologies. Association for Computational Linguistics, Seattle, United States, pp 3886–3893. https://aclanthology.org/2022.naacl-main.285

Personalization Through User Attributes for Transformer-Based Sequential Recommendation

Elisabeth Fischer, Alexander Dallmann, and Andreas Hotho

Abstract Sequential recommendation models are able to learn user preferences based solely on the sequences of past user–item interactions. However, there usually exist additional features about the items as well as the user that can be exploited to improve the model performance. Various models have already successfully exploited additional item information, but as fashion taste, for example, is really dependent on the individual person, characterizing the consumer becomes even more important for recommendations. The integration of user attributes into sequential item recommendation models has received far less attention, although user attributes can alleviate the cold-start problem to some degree through providing general preferences of the user at the start of a session. To address this, we propose a way to fuse user attributes directly at the item representation layer and compare this to a pre- and postfusion baseline. To test our approach, we adapt the well-known model BERT4Rec to use these fused user-item representations as input. We also experiment with a unidirectional transformer model inspired by SASRec. We evaluate all models on a real-world dataset from a fashion e-commerce shop as well as on the public Movielens-1m dataset. Our results show that the integration of user attributes improves the recommendation results for BERT4Rec for next item prediction and can also leverage the user information for recommendations in case of a cold-start setting.

Keywords Sequential recommendation · User attributes · Item recommendation

---

E. Fischer (✉) · A. Dallmann · A. Hotho
Julius-Maximilians University, Wurzburg, Germany
e-mail: elisabeth.fischer@informatik.uni-wuerzburg.de

A. Dallmann
e-mail: dallmann@informatik.uni-wuerzburg.de

A. Hotho
e-mail: hotho@informatik.uni-wuerzburg.de

H. J. Corona Pampín and R. Shirvany (eds.), *Recommender Systems in Fashion and Retail*,
Lecture Notes in Electrical Engineering 981,
https://doi.org/10.1007/978-3-031-22192-7_2

1   Introduction

A successful recommender system needs to learn a user's preferences to give relevant recommendations. A number of neural network architectures have been shown
to learn and model the interest of a user from sequences of user interactions successfully, starting from Recurrent Neural Networks (RNNs) [11] and Convolutional
Neural Networks (CNNs) [21] up to transformer-based models like SASRec [13]
and BERT4Rec [19]. These models are especially useful in recommendation settings where the user is anonymous, as they only rely on the history of previous
interactions and do not require any other information about the user. However, often
additional data characterizing the items or the user is present, which can be explored
to improve the recommender's predictions. The combination of user attributes and
sequence information could help with the cold-start problem, but especially user
attributes have not been explored much yet. For a user returning after a long time
with few or none recent interactions, attributes from the user profile could still produce meaningful recommendations and anonymous users or users without profiles
could still receive recommendations based on their interaction.

The inclusion of additional item information has already been shown to improve
recommendation in several models. For example, in [23] a 3D-CNN is used to learn
a character-wise text-representation of the item and in Novabert [17] embeddings of
categorical item attributes are integrated directly in the transformer layer as key and
query. In contrast, it is hard to find work on the inclusion of user attributes, not the
least because of the lack of publicly available datasets, as releasing user information
is often a privacy issue. However, there is the work of Chen et al. [3], who propose
an unidirectional transformer model using user profile information to increase the
click-through-rate, although the paper does not focus on the user information. They
embed the sequence with a transformer and concatenate an embedding of the user
and other information to the output of the transformer.

While their setup is similar, our goal is the prediction of the next item and our
approach to the problem differs as follows: At each step of the sequence, the user
decides which item to take, and therefore, the representation of the user in our model
should also influence each item separately. Additionally, the user representation and
the sequence contain different types of information, so it would be reasonable to treat
them as different data types.

However, it has been already shown that transformers themselves have the capability to handle different types of data successfully. In VisualBert [16], image and
language representations are used as combined input to the transformer, so the transformer can learn the relationship between both. Combining the user and the sequence
in a similar way could allow the transformer to learn the influence of the user for
each sequence step. Thus, we want to investigate whether the transformer itself can
handle merging the user information into the sequence and compare the performance
to approaches merging the user information before and after the transformer.

We therefore propose a modification for including a representation of the user
for transformer-based recommendation models. First, we create an embedding of

the user's attributes in addition to the usual item embeddings. Then, we concatenate this user embedding to the beginning of the sequence of embedded item IDs. This modified sequence now contains the user embedding as the first entry, followed by each item embedding and is used as an input for the transformer layer, so the attention between each item and the user can be computed.

To test our approach, we adapt the well-known BERT4Rec model as well as an unidirectional transformer model inspired by the SASRec model, but utilizing a cross-entropy loss instead of a negative sampling-based loss function as this loss performs better on our data. We also adapt two baselines: KeBERT4Rec merges the attributes pre-fusion, which means before the transformer layer creates the fused sequence representation. Inspired by [2], we create another model using a post-fusion merge of the attributes on the output of the transformer layer.

We evaluate our models on a proprietary dataset from a large fashion webshop. Additionally, we provide results for Movielens-1m, an publicly available movie recommendation dataset. The Movielens-1m dataset contains user attributes like gender, age and occupation that can be used for sequential recommendation. As both datasets contain no numerical attributes for the user, we limit our evaluation to the use of categorical attributes.

Overall, our contributions are as follows: (1) We propose a way of including categorical user attributes into transformer-based recommendation models with exemplary adaptations for BERT4Rec and a SASRec-inspired model. (2) We evaluate the performance on a real-life dataset from an e-commerce online shop and on the Movielens-1m dataset.

In Sect. 2, we give an overview over the related work for sequential recommendation. We define the task at hand in Sect. 3, followed by the description of our approach in Sect. 4. Our datasets are described in Sect. 5, while we present our experiments and results in Sect. 6. Finally, we conclude the paper in Sect. 7.

2 Related Work

This section covers related work on sequential recommendation with a focus on deep learning methods and architectures using additional information for users and items.

Over time, different neural network architectures have been introduced for modeling sequences of user interactions. GRU4Rec [11] was introduced by Hidasi et al. as the first model for sequential recommendation using Recurrent Neural Networks to model the item sequence and was later improved by introducing a new ranking loss in [10]. However, the authors of [20] show that a cross-entropy loss can perform even better. Other architectures proposed Convolutional Neural Networks, for example Caser [21] or combinations of RNNs and CNNs [26]. Originally introduced for tasks in Natural Language Processing, attention [24] has also been adapted for sequential session recommendation. The first model using attention to encode the sequence was NARM [15], which is based on a GRU layer. In contrast, SASRec [13] uses an unidirectional transformer to encode the sequence and was trained through sampling negative items for a binary cross-entropy loss. Inspired by BERT [7], the authors of [19] adapted the masked training task [22] and bidirectional transformers for sequential recommendation. Their model BERT4Rec was able to outperform state-of-the-art methods on several datasets.

To further improve recommendation performance, several modifications of these models have been published, which try to leverage additional item information. For example, in [23] CNNs are using 3D convolutions to include character-wise item descriptions. In [12], textual and visual item features are fed to different GRUs and output is combined for recommendation afterwards. In [2], the authors use the Latent Cross-Technique to included context in a GRU model in both a pre-fusion and post-fusion setting. For bidirectional transformer models, NovaBert [17] infuses item attribute embeddings directly in the transformer as query and value, while [8] allow the integration of categorical item attributes in BERT4Rec through adding the item embeddings in the embedding layer. The framework Transformers4Rec [18] transfers some more recent NLP models like XLNet [27] for sequential recommendation and allows the integration of categorical and numerical item features. Another type of information about the sequence and the items is explored in MEANTIME [5], where the authors embed the temporal information of the interactions. While these works focus mostly on additional item information, there are some papers taking special interest in the user. CASER [21] for example creates a separate user embedding from the user ids. A similar approach in combination with self-attention is done by Wu et al. [25], but both models only rely on the identity of the user and not on attributes of the user. FairSR [14] includes user attributes to make recommendations more fair and uses a CNN and the construction of a preference graph in a multi-task learning setting to achieve this. In [4] the authors propose a system for next basket recommendation, which models dynamic user attributes.

Most similar to our work is the work of [3], who use an unidirectional transformer and embed user attributes and other features separately from the sequence. The sequence including the target item is encoded with the transformer. The embedded features and the embedded item representation are concatenated and fed through three layers of MLPs for a binary classification of whether the target item is clicked. In contrast, we want to predict the next item, so the target item is not included in our input sequence. We only use the last output of the transformer for our prediction and concatenate the user and the sequence before the transformer layer, as we want investigate the transformers ability to learn the relations between the user and each separate item in the sequence.

3 Problem Setting

In this section, we introduce the problem setting and the notation used in this paper, which closely follows [8]. In this work, we aim to solve the task of recommending items to a user given the sequence of previous item interactions. We define the set of users as $\mathcal{U} = \{u_1, u_2, \ldots, u_{|\mathcal{U}|}\}$ and the item set with $\mathcal{V} = \{v_1, v_2, \ldots, v_{|\mathcal{V}|}\}$.

Therefore, for each user u who interacted at the relative time step t with an item $v_t^u \in \mathcal{V}$, we can denote the sequence of interactions as $s_u = [v_1, v_2 \ldots, v_n]$. The set of all sessions is defined as $\mathcal{S}_u = \{s_{u_1}, s_{u_2}, \ldots, s_{u_{|\mathcal{U}|}}\}$. For each user $u \in \mathcal{U}$, we also have information in the form of categorical attributes which we write as $a_u = \{a_1, a_2, \ldots, a_{|A|}\}$. With $\mathcal{A}$, we denote all possible attributes.

We can now formally define our task as the prediction of the next item v_{n+1} for every $v_t^u \in \mathcal{S}_u$ with the additional user information a_u.

4 User Attributes Personalization for Transformers

In this section, we specify the details for our base models and our original approach leveraging user attributes. In our architectures, we integrate only categorical user attributes into transformer-based models for sequential recommendation. We adapt two well-known recommendation models for our experiments: SASRec [13] as one of the first recommendation models using transformers and BERT4Rec [19] as a bidirectional model with the Cloze training task (Fig. 1).

The original SASRec^{neg} [13] uses a negative sampling loss for training. But in our preliminary experiments, we noticed that the SASRec model was not able to learn much (all metrics smaller than the most popular baseline) for the Fashion dataset. As the cross-entropy loss has worked well for other types of recommendation models (BERT4Rec, GRUs) in [20] and [19], we change the training task: Instead of a binary classification with negative item sampling, we implement a classification of all items with a cross-entropy loss. We denote this architecture with SASRec^{cross} and the original implementation with SASRec^{neg}. We name the architectures including user attributes U-BERT4Rec and U-SASRec^{cross}, respectively, and show them in Fig. 1.[1]

4.1 Layers

For both U-SASRec^{cross} and U-BERT4Rec, the model consists of three different layers, like in their original implementations: (i) the *embedding layer*, which learns a representation of the input (i.e., the sequence of item identifier and the user attributes). The resulting representation is then fed to (ii) a *Transformer layer* that consists of L layers of transformer blocks (as introduced in [24]), which the final output is put through (iii) the *projection layer*, which maps the output of the last transformer back to the item space using a linear layer with softmax activation function.

Embedding Layer The embedding of the interaction sequence $s_u = [v_1, v_2 \ldots, v_n]$ follows the setup in BERT4Rec and SASRec with some modifications for the user. The embedding layer has a size of d and consists of the following parts: (i) the

---

[1] Our code is available at https://github.com/LSX-UniWue/recsys-22-user-attributes-recomme nder.

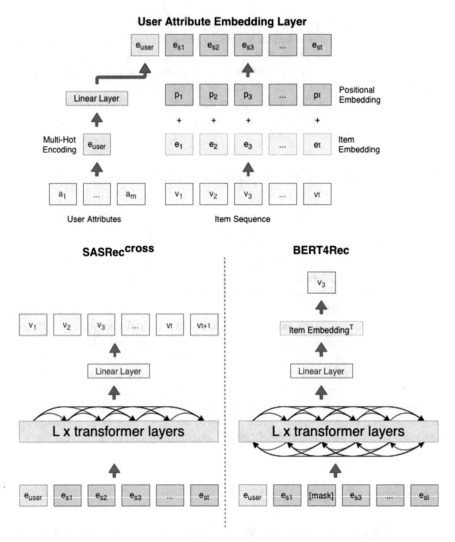

Fig. 1 User attribute embedding layer and the modified architectures for BERT4Rec and SASRec. The user representation is prepended to the item representations in the embedding layer and fed to the transformer layers. For the transformer layers, SASRec contains only the forward connections, while BERT4Rec also contains backward connections

embedding $E_\mathcal{V} \in \mathbb{R}^{|\mathcal{V}| \times d}$ of the item identifier v and (ii) the embedding $E_P \in \mathbb{R}^{N \times d}$ of the timestep t, marking the position of the item in the sequence. This encodes the position for the transformers, with N being the maximum length of the input sequence. (iii) One multi-hot encoding of the user attributes $E_{|A|} \in \{0, 1\}^{|A|}$ and (iv) a linear layer $l_A(x) = Wx + b$ with the weight matrix $W \in \mathbb{R}^{|A| \times d}$ and bias $b \in \mathbb{R}^d$ for scaling the encoded attributes E_A to the hidden size d.

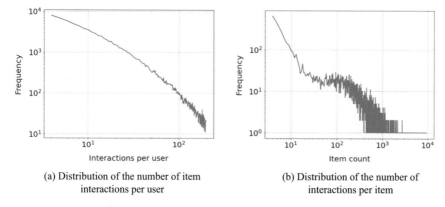

(a) Distribution of the number of item interactions per user

(b) Distribution of the number of interactions per item

Fig. 2 Item and session length distributions for the Fashion dataset

The item embeddings $e_t = v_t E_{\mathcal{V}}$ and the positional embeddings $p_t = t E_P$ are summed up for each time step t. This gives us $e_s = [e_1 + p_1, e_2 + p2, \ldots, e_n + p_n]$ as the list of embedded items. The embedding of a user is created by applying l_A to the multi-hot encoding of a user's attributes, formalized as $e_u = l_A(E_A(a_u))$. We prepend the user embedding e_u to e_s, giving us $e = [e_u, e_{s_1}, e_{s_2}, \ldots, e_{s_n}]$ for the final output of the embedding layer and input for the transformer.

Transformer Layers Each of the L transformer layer contains $N + 1$ transformers (maximum length of the sequence plus one for the user representation) with the hidden size d. In the U-SASRec^{cross} model, the connections between the transformer layers are unidirectional and pointing forwards in the sequence, while in the U-BERT4Rec model they are connected bidirectional.

Projection Layer To map the output of the L-th transformer layer back to the item IDs, a linear layer is used. As the implementation for U-BERT4Rec and U-SASRec^{cross} differs, the details for the projection layer and the training can be found in the following sections.

4.2 BERT4Rec

BERT4Rec uses the Cloze masking task for training. This means the hidden state of the L-th Transformer layer h_t^L at time step t is used to predict the masked item v_t. BERT4Rec uses a linear layer with the GELU activation function [9] and layer normalization, followed by the transposed item embedding $E_{\mathcal{V}}$ to project the output back to the item space. Therefore, we use $o_0 = \sigma(h_t^L W_1)$ with $W_1 \in \mathbb{R}^{d \times d}$ for the linear layer with $\sigma = GELU$, followed by $o_2 = o_1 E_{\mathcal{V}}^T$ for the projection (bias omitted for readability).

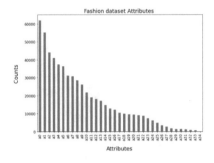 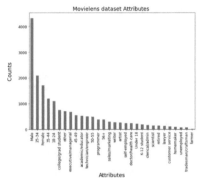

(a) Distribution over all users for gender, age and shopper type attributes for the Fashion dataset

(b) Distribution over all users for gender, age and occupation attributes for the Movielens-1m dataset

Fig. 3 Item and session length distributions for the Fashion dataset

4.3   SASRec^{cross}

As we compute a full ranking over all items instead and use the normal cross-entropy loss to train the model, the projection layer changes as follows:

The hidden state of the L-th Transformer layer h_t^L at time step t is used to predict the item v_{t+1}. To compute the complete ranking over the items, we use the linear projection $o = h_t^L W$ with $\bar{W} \in \mathbb{R}^{d \times |\mathcal{V}|}$.

5   Datasets

In this section, we describe the two datasets used for evaluation. We give information on the preprocessing, some statistics and show the distribution of available user attributes.

As an established dataset for sequential recommendation, we use the Movielens-1m[2] dataset. Movielens-1m contains ratings for movies from an online platform and provides information about the users gender, age and occupation. Unfortunately, none of the larger datasets from Movielens give this information. We apply the same preprocessing steps as BERT4Rec [19]. For training, we use the commonly used leave-out-out split, where the penultimate item in a sequence is used for validation and the final item in a sequence is used for testing the model. The attribute distribution for the Movielens-1m dataset is shown in Fig. 3b.

The second dataset (Fashion) contains user interactions with product pages from a big online fashion store collected over the duration of 45 days. The dataset only

---

[2] https://grouplens.org/datasets/movielens/1m/.

Table 1 Statistics for the train, validation and test split of the Fashion dataset

Dataset	$|\mathcal{U}|$	$|\mathcal{V}|$	$|\mathcal{A}|$	#Interactions (m)	Unknown items (%)
Train	93,248	9560	34	2.4	
Validation	21,651	5770	34	0.4	1.7
Test	46,391	7429	34	1.1	12.6

Table 2 Statistics of the two preprocessed datasets Movielens-1m and Fashion

Dataset	$|\mathcal{U}|$	$|\mathcal{V}|$	$|\mathcal{A}|$	#Interactions (m)	Avg. length	Density (%)
Movielens-1m	6040	3706	18	1	165.6	4.46
Fashion	119,776	10,502	34	3.5	26.4	0.02

contains interactions with pages which are linked to a product, interactions with technical pages (e.g., account pages) and other content pages (e.g., overview) are not included. As for user attributes, we collect information from the following three categories: the gender, the age group and the shopper type. The gender can be one of male, female or other/unknown. The age is also a categorical value, as only a general age group is provided, e.g., users between 20 and 29 years are assigned the same age group. The shopper types define groups of users with specific interests or behaviours. This could be a higher interest in a specific brand or sports or the tendency to buy mostly bargains and products on sale or a mix of different characteristics. Therefore, a user can fall into multiple of these shopper types at once. Figure 3a shows the distribution of user attributes over the complete dataset.

If the same product is interacted with multiple times in a row, we remove all but one interaction—a product can still be in the sequence multiple times, just not immediately after itself. Furthermore, we drop sequences with less than 3 and more than 200 interactions, as in [19]. While the average number of interactions or session length is about 26.4, most of the sessions are on the shorter side, with close to a long tail distribution for long sessions, as shown in Fig. 2a. We show the frequency of the interactions per items in Fig. 2b: There are few items with a high number of interactions, and far more items with a low interaction count. Overall, the density (describing the average number of unique item interactions per user) of the dataset is at 0.02% and far lower than in the Movielens-1m dataset.

We split the dataset by time with 28 days for training, 3 for validation and 14 days for the test set, as shown in Table 1. We can also see that the training item set differs hugely from the items seen in test and validation. Only about 60% of the training items are seen in validation and 1.7% items are new and unknown. In the test period, we see about 66% of the items in the training set, but we also find that 12.6% of the items haven't been seen before. Items not seen in training will be mapped to an special token (<unknown>).

We show the statistics about both the preprocessed Movielens-1m and the Fashion dataset in Table 2.

6 Experiments

In this session, we introduce the setup for our evaluation and present a comparison of our models as well as experiments on the influence of the session length, the performance in a cold-start setting and an ablation study of the attributes.

6.1 Evaluation Setup

To evaluate the influence of the added user representation, we compare the results for both U-BERT4Rec and U-SASRec^{cross} with the models without user information and two baseline models for merging attributes. For evaluation, we report the *Hit Ratio (HR)* and the *Normalized Discounted Cumulative Gain (NDCG)* at $k = 1, 5, 10$. Instead of using sampled metrics, we report the full metrics as it has been shown that sampled metrics are not reliable for ranking the models we're considering in [6]. We also use the one-sided Student's t-test for paired samples to verify the statistical significance of the improvements for α-level 0.01 and 0.05. We train our model with the Adam optimizer and unless specified otherwise use the parameters as in [19].

6.2 User Attribute Baselines

We use two baselines for including user attributes in our model to cover both options of merging information before and after the transformer layer.

6.2.1 KeBERT4Rec

The authors of KeBERT4Rec [8] describe an extension to BERT4Rec for including item keywords in the embedding layer as the transformers input. They create a multi-hot encoding from the keywords for each item, scale it with a linear layer to the dimension d of the transformer and add it element wise to each item embedding to create the transformer input. We follow their approach and add the user attributes as attributes for each item in the sequence instead. Using the notations from Sect. 4, this gives us k_t as the keyword embedding for an item, e_t as the item embedding and p_t as the positional embedding at t. The embedded sequence is then computed as $e_s = [e_1 + p_1 + k_1, e_2 + p2 + k_2, \ldots, e_n + p_n + k_n]$. We set k_t to our user embedding e_u for all timesteps t.

6.2.2 Latent Cross-Post-fusion Context

In [2], the authors introduce the latent cross-technique and successfully integrate context post-fusion, after the sequence has been transformed by the GRU layer. The context is multiplied elementwise to the output of the GRU, serving as a mask. Inspired by this, we adapt a baseline as follows: The output of the transformer layer h_t is modified for each sequence step t to $h_t^* = h_t \odot e_u$, with $\odot$ as the elementwise multiplication. h_t^* is then used as the input for the projection layer.

6.3 Hyperparameters

6.3.1 Fashion Dataset

We select the hyperparameters for the Fashion dataset by running a parameter study with Optuna [1] on the base models without user information. The best parameters are then used to train all models with user attributes. To limit the computation time and resource usage, we set the maximum session length to 50 after first experiments. We sample with a Tree-structured Parzen Estimator from [1, 4] with step size=1 for the transformer heads, [1, 4] and step size = 1 for the number of transformer layers, [0.1, 0.5] and step size = 0.1 for the dropout, [16, 48] and step size = 4 for the hidden size. For the learning rate, we sample log-uniformly between [0.01, 0.0001]. We end up with transformer heads = 2, transformer layers = 2, dropout = 0.1 for both models. For the BERT4Rec models, we use a learning rate of 0.003, a hidden size of 18 and train for 200 epochs. For all SASRec^{cross} models, we end up with a learning rate of 0.001, a hidden size of 20 and train for 100 epochs. We run an additional hyperparameter study for U-BERT4Rec to finetune the model and report the best model as U-BERT4Rec $_{tuned}$. The hidden size is increased to 48 for this model. Due to time limits, we could not conclude further parameter studies.

6.3.2 Movielens-1m Dataset

For the Movielens-1m dataset, we set our hyperparameter settings as in [19] and use transformer heads = 2, transformer layers = 2, dropout = 0.2, batch size = 128, hidden size = 32 accordingly for all experiments and train for 200 epochs.

6.4 Results

In Table 3, we show the results of the base models without additional information for both datasets. We also report the most popular baseline (POP). As expected, the popularity baseline is outperformed by both models by far but the comparison

Table 3 Hitrate and NDCG for BERT4Rec and SASRec^{cross} and the popularity baseline evaluated on the Fashion and Movielens-1m datasets

Dataset	Metric	POP	B4R	SASRec^{cross}
Fashion	HR@1	0.000	0.079	**0.114***
	HR@5	0.003	0.274	**0.279**
	HR@10	0.011	**0.358***	0.356
	NDCG5	0.002	0.180	**0.199***
	NDCG@10	0.004	0.207	**0.225***
Movielens-1m	HR@1	0.003	0.013	**0.051***
	HR@5	0.005	0.050	**0.155***
	HR@10	0.011	0.103	**0.230***
	NDCG5	0.004	0.031	**0.103***
	NDCG@10	0.006	0.043	**0.128***

Significance is marked by *for $p < 0.01$ and +for $p < 0.05$ for the comparison between BERT4Rec and SASRec^{cross}

between both models shows mixed results. On Movielens-1m, the performance of SASRec^{cross} is more than doubled compared to BERT4Rec, but on the Fashion dataset the differences are far smaller. For HR@5, the difference is not significant, and for HR@10, BERT4Rec is even significantly better. Overall, SASRec^{cross} shows itself to be a competitive model to BERT4Rec, with great improvements for the Movielens-1m dataset. Previously, the original SASRec has been reported to perform worse than BERT4Rec, but in [6] SASRec already does perform better for the HR@10 when evaluated on fully computed metrics. Therefore, the performance improvement of SASRec^{cross} on our data is not surprising, but gives reason to explore SASRec and SASRec^{cross} for other datasets in the future.

In Table 4, we show the models leveraging user attributes in comparison with those without.

For the Fashion dataset, we can see significant improvements in HR@1, NDCG@5 and NDCG@10 for the all models using user attributes. The results for HR@5 and HR@10 show slightly worse performance for U-BERT4Rec, but the differences are nonsignificant. However, if we compare the hyperparameter-tuned U-BERT4Rec $_{tuned}$ model to the baseline we see significant improvements over all metrics. The slightly bigger hidden size for this model might give the model the capacity to encode more of the user information. But overall, the improvements for both KeBERT4Rec and LC-BERT4Rec are higher than the ones for U-BERT4Rec and are significant in all cases. The results for the SASRec^{cross} models are mixed: All models perform worse regarding the HR@10, U-SASRec^{cross} even significantly. The highest performance gain is achieved by KeSASRec^{cross}, followed by the LC-SASRec^{cross} model.

On the Movielens-1m dataset, we also get inconclusive results. We see significant improvements for all but HR@1 between BERT4Rec and U-BERT4Rec. For KeBert4Rec and LC-Bert4Rec, only the HR@10 improves significantly, while the other metrics show no improvement or even worse results, even though these are

Table 4 Comparison of NDCG and hitrate between the BERT4Rec-based models with user attributes and between the SASRec^{cross}-based models with user attributes for both datasets

Dataset	Metric	B4R	UB4R	KeB4R	LC-B4R	UB4R$_t$	SAS^c	U-SAS^c	KeSAS^c	LC-SAS^c
Fashion	HR@1	0.0789	0.095*	**0.122***	0.106*	*0.109**	0.114	0.121*	**0.124***	0.123*
	HR@5	0.274	0.273	**0.284***	0.281*	*0.281**	0.279	0.278	**0.282***	0.280
	HR@10	0.358	0.357	0.361*	**0.362***	*0.363**	**0.356**	0.351*	0.355	0.355
	NDCG5	0.180	0.186*	**0.207***	0.196*	*0.198**	0.199	0.204*	**0.207***	0.206*
	NDCG@10	0.207	0.214*	0.231*	**0.222***	*0.225**	0.225	0.227*	**0.231***	0.230*
Ml-1m	HR@1	0.013	**0.014**	0.008	0.011	–	**0.051**	0.046+	0.047+	0.048
	HR@5	0.050	**0.061***	0.051	0.050	–	**0.155**	0.147+	0.150	0.151
	HR@10	0.087	**0.103***	0.102*	0.098*	–	**0.230**	0.226	0.228	0.227
	NDCG@5	0.031	**0.037***	0.030	0.030	–	**0.103**	0.097+	0.099+	0.100
	NDCG@10	0.043	**0.051***	0.046	0.046	–	**0.128**	0.122*	0.124*	0.124

Significance is marked by * for $p < 0.01$ and + for $p < 0.05$ for the different to BERT4Rec and SASRec^{cross}, respectively. U-BERT4Rec is also finetuned for one experiment

nonsignificant. Looking at the SASRec^{cross} models, we see slightly lower results in comparison to the baseline for all models, part of them significant. Only the latent cross-post-fusion model does not show a significant decrease in any of the metrics, but it still performs worse in absolute numbers.

Nevertheless, the general results on the Movielens-1m dataset show the same trend as on the Fashion dataset: Adding user attributes can improve the performance, but between the three different methods of integration attributes, there is no clear winner. The performance increase varies, depending on the chosen dataset and the base model. Fine-tuning a model can increase the performance significantly.

Both the BERT4Rec and the SASRec^{cross} model can profit from the additional user information, with the exception of SASRec^{cross} for the Movielens-1m dataset, where all of the methods perform worse.

Comparing both model types also gives ambiguous results: For the Fashion dataset KeBert4Rec and LC-Bert4Rec perform best, but are outperformed by KeSASRec^{cross} for HR@1. On the Movielens-1m dataset, the performance of SASRec^{cross} is far higher than BERT4Rec. SASRec^{cross} does not profit from the user attributes at all, but is still the best model for the dataset overall. Looking at the different metrics, we can also see that the gains for NDCG tend to be higher than the gains in the hitrate, or that the hitrate even drops. This can be seen for example with the U-BERT4Rec model, but this trade-off between better ranking and fewer hits seems to be a general trend for the models. Due to time and resource limits, we could only conclude the hyperparameter study for U-BERT4Rec on the Fashion dataset and as it shows that we could further improve results, the natural next step would be to conduct further studies for the other models and also for the Movielens-1m dataset.

6.5   Session Length Influence

As we have seen, the inclusion of user attributes can improve the recommendation quality. To understand how the models use the additional information, we take a look at the NDCG and hitrate for the different session lengths for U-BERT4Rec $_{tuned}$ and U-SASRec^{cross} in the Fashion dataset. Figure 4 shows the NDCG@10 and the HR@10 plotted with respect to the length of the input sequence. In Fig. 4a, we see that the performance of U-BERT4Rec is higher than BERT4Rec over most sessions, only at a session length of 40 they perform similar. The number of sessions with greater lengths is relatively small[3] and the better performance for the many shorter sessions manages to increase the overall metrics score. Looking at the hitrate, we see only a slight improvement for shorter sessions, while for longer sessions, the difference is marginalized or even reversed. In Fig. 5, we see a similar pattern in the plots for SASRec^{cross} and U-SASRec^{cross}. For the NDCG, we find a small improvement for sessions up to a length between 35 and 40, but the hitrate decreases slightly over all

---

[3] The number of sessions at the length of 50 is higher, as any longer session will be cut of at 50 and therefore contribute to the count.

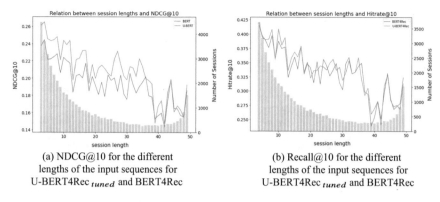

(a) NDCG@10 for the different lengths of the input sequences for U-BERT4Rec $_{tuned}$ and BERT4Rec

(b) Recall@10 for the different lengths of the input sequences for U-BERT4Rec $_{tuned}$ and BERT4Rec

Fig. 4 Metrics shown for each input session length for U-BERT4Rec on the Fashion dataset

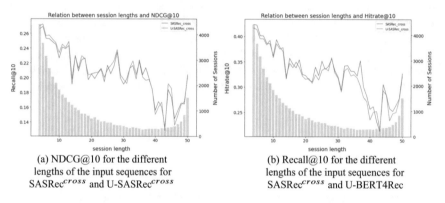

(a) NDCG@10 for the different lengths of the input sequences for SASRec^{cross} and U-SASRec^{cross}

(b) Recall@10 for the different lengths of the input sequences for SASRec^{cross} and U-BERT4Rec

Fig. 5 Metrics shown for each input session length for U-SASRec^{cross} on the Fashion dataset

of the session lengths. The results for Movielens-1m, while not reported here, show a similar pattern in both metrics. Overall, we find that the inclusion of user attributes tends to primarily improve the ranking of the items, especially for shorter sessions. These results show that there is a trade-off between an significantly improved ranking and a slightly worse hitrate for U-BERT4Rec and U-SASRec^{cross}.

6.6 Recommendations Without User Interactions

One advantage of the way U-BERT4Rec and U-SASRec^{cross} integrate the user attributes is the ability to make recommendations even if no interactions are available. To show the performance in a cold-start setting, we remove all item interactions from the Fashion dataset and keep only the user attributes and the target item. We do use one dummy item with the <unknown> token for technical reasons. Table 5 shows the performance for U-BERT4Rec $_{tuned}$ and U-SASRec^{cross} for the modified test data.

Table 5 NDCG and hitrate for U-BERT4Rec and U-SASRec^{cross} on the Fashion dataset in a cold-start setting with no user interactions

Dataset	Metric	POP	UB4R$_{tuned}$	USASRec^{cross}
Fashion users only	HR@1	0.000	0.001	0.001
	HR@5	0.003	0.010	0.005
	HR@10	0.011	0.021	0.011
	NDCG5	0.002	0.005	0.003
	NDCG@10	0.004	0.009	0.005

For both models, the performance drops drastically, as it was to be expected. While the absolute numbers are not high, the metrics are more than doubled in comparison to POP for U-BERT4Rec from e.g., HR@10 of 0.011 to 0.021. The U-SASRec^{cross} model is also beating the most popular baseline, but with less difference. This is in line with the previous results, where the U-SASRec^{cross} model is not able to use the user information as effectively as U-BERT4Rec. Overall, both models perform better than the most popular baseline and can improve recommendations in this cold-start setting.

6.7 Ablation Study

We conclude our experiments with an ablation study of the user attributes, limited to the U-BERT4Rec model due to time restrictions. For each of the attribute categories of age, gender and occupation/shopper type we train the U-BERT4Rec model separately. For the Fashion dataset, we use the parameter of U-BERT4Rec $_{tuned}$ to train the models and also compare to U-BERT4Rec $_{tuned}$. We show the results in Table 6.

On the Movielens-1m dataset, all attributes improve the model's performance slightly, except for the HR@1 metric. The improvements through the gender and age are similar, but for gender, the results are more significant. The occupation has a small, but positive influence on the metrics. This would align with the intuition that an occupation has the least connection to movie preferences, while there are movies targeted at specific age and gender demographics. The combination of all attributes gives the best results for this dataset.

For the Fashion dataset, we see that gender, age and type improve the results significantly over all metrics. Age seems to be the more important attribute, as the performance is higher regarding all metrics. However, the final U-BERT4Rec model with the combination of all three features performs worse than the separate models (except for the HR@10). This shows that the available information is not yet optimally used and that feature selection becomes an important next step with the addition of user attributes.

Table 6 NDCG and hitrate for ablation study of different user attributes for U-BERT4Rec on the Movielens-1m dataset and U-BERT4Rec $_{tuned}$ on the Fashion dataset

Dataset	Metric	B4R	UB4R$_{gender}$	UB4R$_{age}$	UB4R$_{occ|type}$	UB4R$_{(tuned)}$
Fashion	HR@1	0.079	0.117*	**0.121***	0.115*	0.109 *
	HR@5	0.274	0.283*	**0.286***	0.281*	0.281*
	HR@10	0.358	0.362*	**0.363***	0.362*	0.363*
	NDCG5	0.180	0.204*	**0.207***	0.201*	0.198*
	NDCG@10	0.207	0.229*	**0.232***	0.227*	0.225*
Movielens-1m	HR@1	0.013	0.012	0.012	0.011	**0.014** *
	HR@5	0.050	**0.061***	0.058*	0.054	**0.061***
	HR@10	0.087	0.105*	**0.106***	0.102*	0.103*
	NDCG5	0.031	0.036$^+$	0.035	0.033	**0.037** *
	NDCG@10	0.043	0.050*	0.050*	0.048*	**0.051***

Significance is marked by *for $p < 0.01$ and $^+$ for $p < 0.05$

7 Conclusion

While item attributes have been explored to improve sequential item recommendation, the same has not been done for user attributes yet. In this paper, we experiment with the integration of categorical user attributes to transformer-based models. We adapt BERT4Rec and create SASRec^{cross}, a SASRec-inspired model with a cross-entropy loss for our experiments. We propose a model where the user representation is merged to the sequence by the transformers and compare this to pre- and post-fusion approaches. Experiments on two datasets show that the models can profit to different, but significant, extends from user attributes. Especially, the ranking is improved, while the gains in hitrate are lower. We find there is no best approach overall, as the best model is depending on the dataset and base model. Our original approach allows us to give better recommendations for users with no interactions though. Several directions remain to be explored: We can conduct hyperparameter studies to further verify our findings and experiment with different combinations of attributes for a better feature selection. Variants or combinations of the pre- or postfusion approaches could also be explored further.

References

1. Akiba T, Sano S, Yanase T, Ohta T, Koyama M (2019) Optuna: a next-generation hyperparameter optimization framework. In: Proceedings of the 25rd ACM SIGKDD international conference on knowledge discovery and data mining
2. Beutel A, Covington P, Jain S, Xu C, Li J, Gatto V, Chi EH (2018) Latent cross: making use of context in recurrent recommender systems. In: Proceedings of the eleventh ACM international conference on web search and data mining, pp 46–54

3. Chen Q, Zhao H, Li W, Huang P, Ou W (2019) Behavior sequence transformer for e-commerce recommendation in Alibaba. In: Proceedings of the 1st international workshop on deep learning practice for high-dimensional sparse data. ACM. https://doi.org/10.1145/3326937.3341261

4. Chen Y, Li J, Liu C, Li C, Anderle M, McAuley J, Xiong C (2021) Modeling dynamic attributes for next basket recommendation. In: 2022 CARS workshop

5. Cho SM, Park E, Yoo S (2020) Meantime: mixture of attention mechanisms with multi-temporal embeddings for sequential recommendation. In: Fourteenth ACM conference on recommender systems, pp 515–520

6. Dallmann A, Zoller D, Hotho A (2021) A case study on sampling strategies for evaluating neural sequential item recommendation models. In: Fifteenth ACM conference on recommender systems, pp 505–514

7. Devlin J, Chang MW, Lee K, Toutanova K (2019) BERT: pre-training of deep bidirectional transformers for language understanding. In: Proceedings of the 2019 conference of the North American Chapter of the Association for Computational Linguistics: human language technologies, vol 1 (Long and Short Papers). Association for Computational Linguistics, Minneapolis, MN, pp 4171–4186. https://doi.org/10.18653/v1/N19-1423

8. Fischer E, Zoller D, Dallmann A, Hotho A (2020) Integrating keywords into BERT4Rec for sequential recommendation. In: KI 2020: advances in artificial intelligence

9. Hendrycks D, Gimpel K (2016) Gaussian error linear units (GELUs). http://arxiv.org/abs/1606. 08415: Trimmed version of 2016 draft

10. Hidasi B, Karatzoglou A (2018) Recurrent neural networks with top-k gains for session-based recommendations. In: Proceedings of the 27th ACM international conference on information and knowledge management, pp 843–852

11. Hidasi B, Karatzoglou A, Baltrunas L, Tikk D (2016a) Session-based recommendations with recurrent neural networks. In: Bengio Y, LeCun Y (eds) ICLR (Poster)

12. Hidasi B, Quadrana M, Karatzoglou A, Tikk D (2016b) Parallel recurrent neural network architectures for feature-rich session-based recommendations. In: Proceedings of the 10th ACM conference on recommender systems. ACM, New York, NY, USA, RecSys '16, pp 241–248. https://doi.org/10.1145/2959100.2959167

13. Kang WC, McAuley J (2018) Self-attentive sequential recommendation. In: 2018 IEEE international conference on data mining (ICDM). IEEE, pp 197–206

14. Li CT, Hsu C, Zhang Y (2022) FairSR: fairness-aware sequential recommendation through multi-task learning with preference graph embeddings. ACM Trans Intell Syst Technol 13(1). https://doi.org/10.1145/3495163

15. Li J, Ren P, Chen Z, Ren Z, Lian T, Ma J (2017) Neural attentive session-based recommendation. In: Proceedings of the 2017 ACM on conference on information and knowledge management, pp 1419–1428

16. Li LH, Yatskar M, Yin D, Hsieh CJ, Chang KW (2019) Visualbert: a simple and performant baseline for vision and language. 1908.03557

17. Liu C, Li X, Cai G, Dong Z, Zhu H, Shang L (2021) Non-invasive self-attention for side information fusion in sequential recommendation. arXiv preprint arXiv:2103.03578

18. de Souza Pereira Moreira G, Rabhi S, Lee JM, Ak R, Oldridge E (2021) Transformers4Rec: bridging the gap between NLP and sequential/session-based recommendation. In: Fifteenth ACM conference on recommender systems, pp 143–153

19. Sun F, Liu J, Wu J, Pei C, Lin X, Ou W, Jiang P (2019) BERT4Rec: sequential recommendation with bidirectional encoder representations from transformer. In: Proceedings of the 28th ACM international conference on information and knowledge management—CIKM 19. ACM Press. https://doi.org/10.1145/3357384.3357895

20. Tan YK, Xu X, Liu Y (2016) Improved recurrent neural networks for session-based recommendations. In: Proceedings of the 1st workshop on deep learning for recommender systems, pp 17–22

21. Tang J, Wang K (2018) Personalized top-n sequential recommendation via convolutional sequence embedding. In: Proceedings of the eleventh ACM international conference on web search and data mining, pp 565–573

22. Taylor WL (1953) " Cloze procedure": a new tool for measuring readability. J Mass Commun Q 30:415–433
23. Tuan TX, Phuong TM (2017) 3D convolutional networks for session-based recommendation with content features. In: Proceedings of the eleventh ACM conference on recommender systems. ACM, New York, NY, USA, RecSys '17, pp 138–146. https://doi.org/10.1145/3109859.3109900
24. Vaswani A, Shazeer N, Parmar N, Uszkoreit J, Jones L, Gomez AN, Kaiser L, Polosukhin I (2017) Attention is all you need. CoRR abs/1706.03762, http://arxiv.org/abs/1706.03762
25. Wu L, Li S, Hsieh CJ, Sharpnack J (2020) SSE-PT: sequential recommendation via personalized transformer. In: Fourteenth ACM conference on recommender systems, pp 328–337
26. Xu C, Zhao P, Liu Y, Xu J, SSheng VS, Cui Z, Zhou X, Xiong H (2019) Recurrent convolutional neural network for sequential recommendation. In: The world wide web conference on—WWW'19. ACM Press. https://doi.org/10.1145/3308558.3313408
27. Yang Z, Dai Z, Yang Y, Carbonell J, Salakhutdinov RR, Le QV (2019) Xlnet: generalized autoregressive pretraining for language understanding. In: Advances in neural information processing systems 32

Reusable Self-attention-Based Recommender System for Fashion

Marjan Celikik, Jacek Wasilewski, Sahar Mbarek, Pablo Celayes, Pierre Gagliardi, Duy Pham, Nour Karessli, and Ana Peleteiro Ramallo

Abstract A large number of empirical studies on applying self-attention models in the domain of recommender systems are based on offline evaluation and metrics computed on standardized datasets, without insights on how these models perform in real-life scenarios. Moreover, many of them do not consider information such as item and customer metadata, although deep learning recommenders live up to their full potential only when numerous features of heterogeneous types are included. Also, typically recommendation models are designed to serve well only a single use case, which increases modeling complexity and maintenance costs, and may lead to inconsistent customer experience. In this work, we present a reusable attention-based fashion recommendation algorithm (AFRA) that utilizes various interaction types with different fashion entities such as items (e.g., shirt), outfits and influencers, and their heterogeneous features. Moreover, we leverage temporal and contextual information to address both short- and long-term customer preferences. We show

M. Celikik (✉) · J. Wasilewski · S. Mbarek · P. Celayes · P. Gagliardi · D. Pham · N. Karessli · A. P. Ramallo
Zalando SE, Berlin, Germany
e-mail: marjan.celikik@zalando.de

J. Wasilewski
e-mail: jacek.wasilewski@zalando.de

S. Mbarek
e-mail: sahar.mbarek@zalando.de

P. Celayes
e-mail: pablo.celayes@zalando.de

P. Gagliardi
e-mail: pierre.gagliardi@zalando.de

D. Pham
e-mail: duy.pham@zalando.de

N. Karessli
e-mail: nour.karessli@zalando.de

A. P. Ramallo
e-mail: ana.peleteiro.ramallo@zalando.de

© The Author(s), under exclusive license to Springer Nature Switzerland AG 2023
H. J. Corona Pampín and R. Shirvany (eds.), *Recommender Systems in Fashion and Retail*,
Lecture Notes in Electrical Engineering 981,
https://doi.org/10.1007/978-3-031-22192-7_3

its effectiveness on outfit recommendation use cases, in particular: (1) personalized ranked feed; (2) outfit recommendations by style; (3) similar item recommendation, and (4) in-session recommendations inspired by most recent customer actions. We present both offline and online experimental results demonstrating substantial improvements in customer retention and engagement.

Keywords Recommender systems · Transformers · Fashion outfits

1 Introduction

Fashion contributes to our everyday life in reflecting personality, culture and style. With the vast choice of items available in e-commerce, it has become increasingly difficult for customers to find relevant content, combine it and match with a specific style. Finding inspiration, such as for example influencers, has become crucial to support customers in discovering inspirational and fresh outfits tuned to their taste.

Zalando is a leading European online platform for fashion and lifestyle, and one of its main goals is to provide fashion inspiration and support customers in their fashion discovery journey. We do this through different experiences, such as for example creator outfit inspirational content, in touch points such as: (1) "Get the Look" (GTL), a personalized feed of all available influencer outfits (Fig. 1a); (2) "Style" carousel, providing personalized outfit recommendations in different styles (e.g., classic, casual, streetwear; Fig. 1b); (3) "Inspired by you" carousel, providing real-time in-session recommendation based on the customer's recent interactions (Fig. 1c); and (4) "You might also like" carousel, providing personalized recommendations related to the currently viewed outfit (Fig. 1d).

A complexity that arises when building recommendation models for inspirational content in the e-commerce domain is the availability of input signals. Interaction data with non-product fashion entities, such as influencer outfits, is usually low, which makes the data extremely sparse when compared to interactions with individual items (e.g., a shirt). In this work, we investigate how interactions with other "high-traffic" fashion entities could be used and improve "low-traffic" entities such as outfit recommendations.

In e-commerce, often different but related customer experiences are served by completely different recommender systems, for example either to steer more to in-session recommendations or to capture longer-term customer preferences. This can significantly increase not only modeling complexity and maintenance costs of those different systems, but also lead to inconsistent customer experience. In contrast to the common belief that different recommenders are needed for different use cases [24], we show that a single transformer-based recommender system can be trained on diverse types of interactions coming from various sources and successfully re-used across many related use cases, such as session-based recommendation for short-term interests, personalized ranking based on long-term user preferences and even provide item-related or similar recommendations.

(b) "Style preview" carousel with personalized outfits selection in different styles

(a) "Get the Look" personalized feed (catalog) of ranked outfits

(c) "Inspired by you" carousel with in-session outfit recommendations based on customer's recent interactions

(d) "You might also like" carousel with personalized outfit recommendations (right) related to an anchor outfit (left)

Fig. 1 Different outfit recommendation and ranking use cases

Moreover, in e-commerce, many users are either new or do not login when navigating; thus, it is important to use a model that can provide meaningful recommendations to new, cold-start users. The model we propose can serve "partial" cold-start users (users that have not interacted with outfits but have interacted with the platform) by utilizing other types of interactions in the session, as well as "full" cold-start users (without any platform activity) by utilizing contextual information about the user and the use case, such as premise (site or product), device, country, language, etc.

In recent years, sequential recommendation approaches have gained significant traction [10]. These model customers as sequences of interactions to better infer context and customer preferences that change over time. In addition, representing customers as sequences of actions at least partially simplifies the feature engineering and modeling efforts. Sequential recommendation approaches often employ NLP techniques such as Word2Vec [16], RNNs [11, 19] and attention mechanisms [5, 12, 25]. Transformer-based models are the most promising sequential and session-based models to date due to their ability to model long-range sequences and their scalability thanks to efficient parallel training [12, 23, 25]. However, the majority of previous studies are based on offline evaluation and metrics computed on open datasets, which may not be directly comparable with industrial datasets and online results. Moreover, many of them do not consider side information such as categorical inputs to represent items or customers, although it is well known that deep learning recommender systems live up to their full potential only when numerous features of heterogeneous types are used [24]. Due to these shortcomings, even if relevant, most of the research work on self-attention models for recommendations does not explore the potential effectiveness in real-world industry applications. In this work, we contribute to bridge this gap by presenting online experimental results of using a self-attention-based recommender system that can utilize contextual information about the customer and the use case. We showcase that our personalized outfit recommendation model can improve engagement and customer retention in different use cases throughout the customer journey.

In summary, the main contributions of this work are as follows: (1) a reusable transformer-based recommender system for fashion recommendations that utilizes diverse types of interactions on various fashion entities. It is able to provide session-based recommendations as well as take into consideration long-term user preferences and contextual information, to both recurring and cold-start users; (2) extensive A/B testing that shows that this approach can be successfully applied to different recommendation use cases including personalized ranked feed, outfit recommendations by style, similar item recommendation, and in-session recommendations inspired by most recent customer activities; (3) extensive offline experiments and evaluation considering different ranking losses and metrics where we show how our approach substantially increases customer engagement and retention. The rest of the paper is structured as follows: in Sect. 2, we discuss related work; Sect. 3 describes the proposed approach; we present offline evaluation in Sect. 4 and online experiment results in Sect. 5; finally, we provide discussion of conclusions and future work in Sect. 6.

2   Related Work

The main objective of recommender systems is to recommend relevant content tailored to the customer's preferences and intent. Fashion e-commerce has heavily invested in developing recommender systems for different use cases to aid online shopping experience [4, 7] including recommending relevant items [2, 8, 35], complete outfits [3, 9, 14], and size recommendation [13, 22].

Transformer-based recommendation systems started with the SASRec algorithm presented in [12], where similarly to GPT [20], a transformer encoder with masked attention for causal language modeling (CLM) was used to predict the next item in the sequence of user interaction. A similar work in [30] builds on this idea by concatenating user embeddings to each learned item embedding from the same sequence to add contextual user information. BERT4Rec [25] is another work that employs the transformer encoder trained with the masked language modeling (MLM) approach, while masking only the last item of the sequence during inference to avoid leaks of future information.

There are a few works in the literature that use complex hierarchical architectures or mix together different network architectures. The SDM algorithm introduced in [15] is fusing LSTMs and self-attention networks to obtain a final user behavior vector. Ying et al. [32] proposed a similar approach by using a two-layer hierarchical attention-based network while [31] employs a graph and a self-attention network. In all of these approaches, no evidence is provided as to why the complex architecture is needed. Moreover, these methods are only compared to baselines that are either not based on self-attention or not based on deep learning methods. Hence, it is unclear whether the improvement comes from the choice of the architecture. In our case, we present an effective and simple approach that is able to capture both short-term and long-term user interests.

Most of the published studies do not consider side information for representing items or users and instead work with user-item interaction data only. The few works that include side information usually consider only a few item features such as category and brand [15]. The BST algorithm from Alibaba [5] considers richer item features and user profiles as well as contextual features. The differences to our approach are twofold: first, the contextual features are not part of the self-attention mechanism but rather concatenated with the output of the transformer, and second, the use of a positional encoding for the entire sequence instead of representing different sessions separately.

There are only a few lines of work that we are aware of that consider (re)using models for multiple recommendation use cases. Parallel to our work, the work in [29] shows that RNNs can perform well in both, short- and long-term recommendation tasks when certain improvements are applied and even over-perform more complex hierarchical models. The work [34] presents a multi-graph structured multi-use case recommendation solution which encapsulates interaction data across various use cases and demonstrates increase in CTR and video views per user. Another work in [33] surveys the benefits of pre-training and knowledge transfers in recommender systems in order to alleviate the data sparsity problem.

3 Algorithm

3.1 Problem Formulation

Let $\mathcal{U} = \{u_1, u_2, ..., u_{|\mathcal{U}|}\}$ denote the set of users and we denote a set of items by $\mathcal{V} = \{v_1, v_2, ..., v_{|\mathcal{V}|}\}$, we represent the interaction sequence as $\mathcal{S}_u = \{v_1^{(u)}, ..., v_t^{(u)}, ..., v_{n_u}^{(u)}\}$ with interactions ordered chronologically for the user $u \in \mathcal{U}$, where $v_t^{(u)} \in \mathcal{V}$ is the item that the user u has interacted with at time t, and n_u is the length of interaction sequence for user u. Let $\mathcal{C}_u$ be the contextual information about user u, such as country, device, premise and date. Given the interaction history $\mathcal{S}_u$ and context $\mathcal{C}_u$, sequential recommender system aims to predict the item that user u is most likely to interact with at time $n_u + 1$, i.e., we would like to find an item i so that the probability $p\left(v_{n_u+1}^{(u)} = i | \mathcal{S}_u, \mathcal{C}_u\right)$ is maximized.

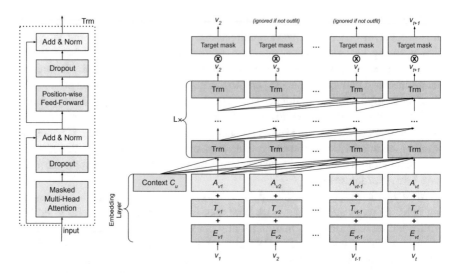

Fig. 2 Illustration of a single transformer encoder (left) and overview of our model architecture (right). Model inputs consist of contextual information and a sequence of: item embedding E_{v_t}, action embedding A_{v_t} and session embedding T_{v_t}. Target mask filters predictions of entities other than the target ones—in this example target mask is set for outfit entities

3.2 Model Architecture

We use a standard transformer encoder [12, 23, 28] trained with the causal language model (CLM) approach as illustrated in Fig. 2. A causal attention mask is provided to the self-attention mechanism so that each position can attend only to earlier positions in the sequence. Note that our approach is oblivious of the training logic, and it can be trained with the masked language model (MLM) approach as well [23, 25]. Given an input sequence of length t and a matrix of learnable input embeddings $V \in \mathbb{R}^{t \times d}$, a single layer of the transformer model computes hidden representations $H^1 = \text{Trm}(V) \in \mathbb{R}^{t \times d}$. By stacking multiple layers, we have $H^l = \text{Trm}(H^{l-1})$. For details, we refer the reader to [12, 28]. The final output of the last position is fed into a softmax layer over the item set. We employ categorical cross-entropy as a loss function, however, experiment with other losses as well.

Given an input sequence, the expected target output is a shifted version of the same sequence. In our setting, an item can refer to a fashion article, an outfit or an influencer. We train on every item in a sequence, but predict only those items that are relevant for the use case. To that end, we assign a Boolean mask to each relevant item. The mask is set to 1 only if the next position in the sequence is associated with an item that should be predicted by the model as a valid recommendation. For example, if our model recommends outfits, then all inputs corresponding to valid outfits will have a mask of 1, for other entities the mask will be set to 0 (all use cases presented in Fig. 1 consider outfits, so in our experiments we focus on outfit predictions). The mask is passed to the loss function and the positions corresponding to zeroes do not contribute to the loss. In addition, items that are valid but not available (e.g., out of stock), will have an output mask set to 0 as well.

Contextual information about the use case and the customer are encoded as embeddings with the same dimensionality as the input item embeddings and are set as the first positions of the sequence so that every other position can attend and utilize this information when making predictions. Inputs such as location, market and language play an important role for cold-start customers that are new to the platform and do not have any interactions yet. For cold-start customers that are new to the outfit use cases but not new to the platform itself, the model can make recommendations based on interactions with other fashion entities, either historical or from the current session. Inputs such as premise and device help the model to hone in on the particular use case. Figure 3 provides a summary of the modeling choices and the different sources of data used for training.

3.3   Input Embeddings

We represent each item v_t (for brevity we omit the superscript) in a user sequence $\mathcal{S}_u$ as a concatenation of learned embeddings that correspond to its different features. The features are always encoded in the same order. Since each input position in the sequence must consist of the same embeddings, we pad the corresponding part of the input tensor with zeroes if the item does not have a certain feature. For example, an outfit might have an influencer (creator); however, a single article does not have one. An item v_t will have the following representation:

$$E_{v_t} = E_{f_1}(v_t) \oplus E_{f_2}(v_t) \oplus \cdots \oplus E_{f_n}(v_t)$$

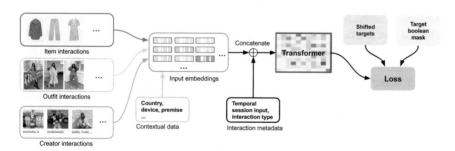

Fig. 3 Overview of the model trained on interactions coming from different fashion entities. Interactions are converted into embeddings of the same structure in terms of feature representation. Contextual features corresponding to the use case such as premise (site) and device are added to input sequences, where temporal session data and interaction metadata is concatenated directly to the learned embeddings. Masking is used in order to consider target outputs only of certain entity types (e.g., outfits)

where E_{f_k} is the embedding matrix of a feature f_k and $\oplus$ is the concatenation operator. Depending on the feature type and its cardinality, we represent features through learned embedding matrices or by using 1-hot encoded vectors. Note that if v_t does not have the feature f_k, $E_{f_k}(v_t)$ is the 0-vector. Examples of categorical features that we represent as embedding matrices are brand, category, colour and influencer, while features such as average price bucket and outfit style are represented by using 1-hot encoding.

As there are clear hierarchical relationships between fashion entities, e.g., an outfit consists of articles and influencers can be represented by the set of outfits they have created; we employ a simple representation of more complex items by averaging the embeddings of the individual items. For example, given an outfit o, let E_B be an embedding matrix for the brand feature. We represent the brand embedding of outfit o as

$$E_B(o) = \sum_{a \in o} E_B(a)$$

where a is an article in o. We represent individual articles in a sequence as outfits with a single article.

For the model to be able to utilize the interaction type of an item and learn to treat different interactions differently, we concatenate a one-hot encoding of the interaction type of the item with the item representation, i.e., $E_{v_t} \oplus A_{v_t}$, where A_{v_t} is one-hot encoding of the v_t's interaction type.

3.4   Modeling Sessions for Long and Short-Term Interests

In this section, we present a simple but yet effective mechanism that allows our algorithm to utilize information from the current and past browsing sessions in order to model short in-session and long-term customer preferences. A user sequence S_u consists of different browsing sessions. A session is a list of consecutive interactions within a given time frame, for example a day, when the user has a clear shopping intent while their interests can change sharply when they start a new session, say a week later. Hence, modeling user sequences directly while ignoring this structure will affect performance negatively (as observed in our experiments).

We model sessions via introducing temporal inputs in the form of interaction recency defined as number of days passed since the action has taken place relative to the model training timestamp during training and the serving timestamp during serving. We discretize recency and consider only the integer part of the timestamp. Each temporal input is assigned its own interaction recency that is concatenated with the rest of the item embeddings. For simplicity, we consider user activity during a single day as a single session. Hence, interactions within the same session will have the same recency. For example, if the next prediction is in-session, then the recency of the last action will be 0. However, the model is able to attend to previous actions as well and use the attention mechanism to select actions that are relevant for the prediction.

Note that with CLM training, the model does not have bidirectional access to positions in the sequence, and hence, it has no information about when the next action has taken place. However, this information is needed for a correct prediction. Namely, the next action may happen on the same day or a week after the previous action. To account for this, in addition to recency, we include the time gap between two actions and make this information accessible to the previous action as one of its inputs. At inference time, the time gap is always 0. Hence, the final input representation of a single interaction to the model is

$$E_{v_t} \oplus T_{v_t} \oplus A_{v_t}$$

where E_{v_t} is the item embedding, A_{v_t} is the action embedding for item v_t, and $T_{v_t} = r_{v_t} \oplus t_{v_t}$ is the session embedding of item v_t, r_{v_t} is the learned embedding of the discretized recency and t_{v_t} is the time gap between interactions with item v_t and item v_{t+1}.

4 Offline Evaluation

4.1 Dataset

Our data consists of a sample of 60 days of user interactions on the platform. We perform a time-based split, where we train on the first 59 days of data and evaluate on the last day. We aggregate the interaction per user into sequences. During the evaluation, we feed only those interactions as inputs that took place before the split timestamp (if any). We filter out sequences without outfit interactions. Our training data contains roughly 6.6m sequences on 9.9k distinct outfits created by 677 unique creators. The test data contains roughly 172k interactions on 9.4k distinct outfits. The average outfit length was 4.6 articles. The total number of distinct articles in the dataset is 1.6m. Roughly, 23% of the customers are completely new to the platform, 31% new to outfits but have interactions with articles on the platform and 46% of the customers have interactions with outfits. To confirm the consistency of our results, we set fixed random seeds, train and evaluate on 3 different datasets (collected with respect to 3 different last days) and report the average performance.

4.2 Experimental Setting

We evaluate our new attention-based fashion recommendation algorithm (AFRA) against a set of existing recommendation algorithms some of which have been powering outfit recommendation use cases presented in Fig. 1. In the following, we present each of the compared algorithms.

Neural LTR: powers the "Get the Look" personalized feed and the "Style" carousel use cases. It is a neural Learning-To-Rank approach implemented by using the TensorFlow-Ranking [18] library to rank outfits based on outfit and user inputs. The outfit inputs consist of learned embeddings of categorical features such as brand, influencer, color and the market of the outfit. The user features are represented by the top-3 favorites brand, colors, influencers, market as well as historical normalized frequencies of how often the customer interacted with certain content types. A feed-forward network is used as the model architecture, with 2 dense layers with 128 and 64 layers units each. The output is passed through a *tanh* activation function. A pairwise logistic loss is used for training. For the "Style" carousel use case, the output is further filtered by style.

IB-CNN-kNN: is a kNN-based recommendation algorithm powering the "Inspired by you" carousel. Outfit recommendations are retrieved based on cosine similarity between item embeddings of previously interacted fashion articles and item embeddings of articles in the outfit. The similarity is defined as the average cosine similarity between the embeddings of the best matching article from the user's history, considering the last 10 article interactions in near real-time and the 200 most recent outfits on the platform. Each item is represented by latent embeddings computed by using a fine-tuned CNN as in [1].

IB-CF-kNN: is a kNN-based Collaborative-Filtering algorithm powering the "You might also like" carousel. We retrieve outfit recommendations based on cosine similarity between user-outfit vectors.

AFRA: AFRA has been implemented as described in Sect. 3 by using the CLM training approach, 2 transformer layers each with 8 attention heads, dff layer with 1024 units, dropout of 0.1, d_{model} set to 128, batch size of 64 and learning rate set to 0.01. The algorithm was trained with 10 epochs. Each interaction sequence passed to the model has been truncated to the 100 most recent interactions. The model uses a set of article-based categorical features such as brand, color, material, fit, pattern; outfit based-features such as influencer and style, general interaction features such interaction type and interaction recency and contextual features such as premise, market and device type. We employ two strategies to evaluate AFRA: **AFRA-RT** in which the customer actions are available to the algorithm in near real-time and **AFRA-Batch** in which only customer actions performed up to the previous day are available to the algorithm.

SASRec: SASRec [12] is a transformer-based recommendation algorithm trained with the CLM approach. We apply the same hyper-parameters that we use in AFRA. We experiment with the authors' implementation of the algorithm. The model uses only embedding representation of user and item IDs as inputs. The model is trained on both outfit and article interactions, while predictions are made on outfits only.

It is worth noting that AFRA is able to make use of additional content, contextual or temporal features that other approaches reasonably cannot (CF, kNN, LTR) or do not (SASRec) use without introducing substantial changes in the algorithms.

4.3   Offline Experiments

In this section, we evaluate different aspects of our proposed algorithm such as *relevance*, *diversity* and *freshness*. We experiment by using different variants of AFRA by training on different types of fashion items. In addition, we evaluate the performance of AFRA by using other ranking loss functions that are commonly considered in the literature.

4.3.1   Relevance

We run offline evaluation using a number of relevance metrics, including recall@k, precision@k, hitrate@k, nDCG@k and mAP@k according to their standard definition provided in [27]. For each algorithm, we calculate the metrics for the top k results, where $k \in \{5, 15, 30\}$ which corresponds to the use cases that we consider, e.g., $k = 15$ and $k = 30$ corresponds to the first and the second page of results in a feed, while $k = 5$ corresponds to a single carousel of recommendations. The recommendation task consists of predicting outfit clicks, regardless whether they are sequential (i.e., in-session), historical (i.e., when customers return from another day) or cold-start (when customers perform an action for the first time). Since all of the relevance metrics are strongly correlated with each other, for brevity we report only recall@k, which in turn correlates well with click-through-rate (CTR) [11].

Table 1 shows the recall@k for all the compared algorithms. The following main observations can be made. First, the two versions of AFRA that simulate real-time and batch scenarios perform better than all other algorithms, including SASRec, on all and on cold-start customers specifically. Second, training by using diverse item interaction data substantially improves the relevance compared to training on outfits only. This can be observed on both algorithms, AFRA and SASRec. Third, AFRA performs substantially better than SASRec even when trained on the much sparser outfit interactions only, thanks to using rich heterogeneous data and session encoding. Fourth, AFRA performs substantially better on cold-start customers compared to all other algorithms thanks to its ability to utilize contextual data. This is especially pronounced in the real-time use case where session data is available to the algorithm. Fifth, using real-time session data has a large impact on performance: in certain scenarios AFRA-RT achieves twice as high a recall compared to AFRA-Batch.

4.3.2   Freshness

Recommender systems have bias toward older items since those usually have more interactions. Users, however, prefer fresh content, although not at the expense of relevance [6]. We focus on freshness in the "Get the Look" feed use case, which is our main entry point for fashion inspirations. We measure freshness of recommendations as the average age of the top-30 recommended outfits, given in days. Table 3

Table 1 Relevance in terms of recall@k for all compared algorithms computed on new and on all customers

Customer segment	Algorithm	Recall@5	Recall@15	Recall@30
All customers	AFRA-RT	**0.135**	**0.233**	**0.301**
	AFRA-Batch	0.081	0.156	0.224
	AFRA-RT (outfits only)	0.093	0.158	0.214
	AFRA-Batch (outfits only)	0.056	0.108	0.156
	LTR	0.025	0.104	0.153
	IB-CNN-kNN	0.056	0.078	0.110
	IB-CF-kNN	0.052	0.073	0.102
	SASRec	0.061	0.102	0.144
	SASRec (outfits only)	0.051	0.075	0.099
New customers (cold-start)	AFRA-RT	**0.082**	**0.164**	**0.232**
	AFRA-Batch	0.045	0.102	0.163
	AFRA-RT (outfits only)	0.069	0.131	0.183
	AFRA-Batch (outfits only)	0.045	0.093	0.134
	LTR (w/recency)	0.033	0.085	0.122
	Popularity	0.030	0.052	0.091
	IB-CNN-kNN	0.034	0.042	0.055
	IB-CF-kNN	0.047	0.057	0.074
	SASRec	0.041	0.074	0.101
	SASRec (outfits only)	0.005	0.011	0.032

It can be observed that all AFRA variants perform the best in general for the given use cases. The label "outfits only" means that the model has been trained by using outfit interactions only
Bold only highlights the best value

Table 2 Change in AFRA's recall@k when various loss functions are used compared to full cross entropy loss

Loss	Recall@5 Δ (%)	Recall@30 Δ (%)
Sampled cross-entropy	−19	−13
Binary cross-entropy [12]	**−7**	**0**
BPR	−32	−11
TOP1	−73	−20

In summary, the gain in training speed does not justify the drop in the offline evaluation metrics
Bold only highlights the best value

Table 3 Average outfit age in the top-30 recommendations when different freshness strategies are used

Algorithm	Freshness@30
LTR	53
AFRA	51
+ Age feature	40
+ Age decay	**20**

Bold only highlights the best value

Table 4 A/B test results showing increase in user retention and engagement metrics of our transformer-based recommender system, compared against the existing algorithms for three use cases illustrated in Fig. 1

Customer segment	Get the look		Style preview		Inspired by you	
	Retention (%)	Engagement (%)	Retention (%)	Engagement (%)	Retention (%)	Engagement (%)
All customers	+28.5	+33.1	+23.9	+30.2	+130.5	+201.3
New customers (cold-start)	+42.0	+47.1	+39.7	+39.7	+109.2	+137.5
Existing customers	+27.0	+32.0	+23.1	+29.6	+130.5	+201.9

shows this metric for the top-30 recommendations. We can observe that the previous approach, LTR and AFRA provide similar freshness. Moreover, we have conducted experiments on how this metric could be improved, and considered two strategies for adjusting for freshness. The first one is inspired by [6], where an item age feature is added during training which is set to 0 during inference to "de-bias" old outfits that have higher chance of being interacted with. The second one introduces a tuned age exponential decay, simulating "content aging", with a half-life of 3 weeks obtained by parameter tuning. This value is used to weight the ranking score produced by the model during inference, which can be seen as a re-ranking of the results. Exponential decay strategy proved to be particularly effective as it does not harm relevance and substantially increase fresh content among the top-k recommendations, decreasing the average age from 51 to 20 days. On the other hand, the age feature strategy decreased the average age to 40 days (Table 3).

4.3.3 Diversity

Diversity is another aspect of recommender systems important to prevent filter bubbles that cause the customers to lose interest over time due to recommendations that are too similar [26]. We measure two types of diversity: *inter-list diversity* (measuring the content diversity within a list of recommendations) and *temporal diversity* (measuring the difference in recommendations from one to the next visit). As a proxy for inter-list diversity we use the maximum consecutive sublist in the top-k created by the same creator (consecutive recommendations from the same creator are undesirable for our algorithm). AFRA and LTR had the highest diversity among all algorithms. For both algorithms, this metric on average is less than 2.0, with AFRA outperforming LTR by up to 20%. We define temporal diversity as a normalized set difference between the recommendation lists from two consecutive visits. Both algorithms perform similarly with temporal diversity roughly around 70%. As a future work, we would like to introduce impression data to AFRA to improve temporal diversity by down-ranking items, the user has already seen but not interacted with.

4.3.4 Loss Functions

We employ other standard ranking functions based on negative sampling such as BPR [21], TOP1 [19] and binary cross-entropy [12] to improve relevance and/or training speed. Table 2 shows the relative change in recall compared to standard categorical cross-entropy (softmax loss). We experiment with using 30 and 100 negative samples (without replacement). The main observation is that loss functions based on negative sampling are not very effective in our setting. The training speed improvements obtained are modest and always less than 2x. Hence, the decrease in relevance does not justify the improvements in training speed. One of the possible reasons for decreased relevance metrics could be the choice of negative samples not matching the background distribution well. Improving the negative sampling distribution as well as using hard negatives are among our future work directions.

5 Online Results

In order to confirm the efficacy of our algorithm in real scenarios, we have performed A/B tests on three of our use cases: "Get the Look", "Style preview" and "Inspired by you" (Fig. 1a–c, respectively). Each A/B test was run for 3–4 weeks (until convergence). For both "Get the Look" and "Style preview", we compare AFRA-Batch against LTR in order to be consistent with the previous approaches that used daily updates of customer interaction data. Real-time session data is used only for new customers to address the cold-start problem. For the "Inspired by you" use case, we use near real-time data and therefore compare AFRA-RT against the IB-CNN-kNN algorithm that uses near real-time data as well. Table 4 summarizes the results from the A/B tests on the *retention* and *engagement* KPIs. We define retention as the share of users with multiple interactions within 7 days, and engagement as outfit interaction rate per user.

In summary, AFRA performs substantially better on all KPIs on all tested use cases. We can observe that on the "Get the Look" and "Style preview" use cases the improvements in retention range from 23% up to 42%. The improvement was even higher on the "Inspired by you" use case where it ranged from 109 to 130%. We observe a strong improvement on the engagement KPI as well that ranges from 30 to 47% for the first two use cases, reaching 201% on "Inspired by you". On both KPIs, the improvements are stronger on the cold-start customers on the first two use cases, thanks to introducing in-session recommendations and contextual inputs in AFRA.

These strong results are consistent with our offline experiments. We believe this is because we train on diverse sources of interaction data from all products and premises which in turn helps to provide more personalized and relevant content and dampens feedback loops and selection biases [17, 24].

6 Conclusions

In this paper, we have presented a reusable transformer-based recommender system that is able to utilize different types of interactions with various fashion entities. We have shown our approach is able to model short-term customer interests by providing session-based recommendations as well as take into consideration long-term user preferences and contextual information about the customer and the use case. We have demonstrated its effectiveness on different use cases, with extensive offline and online experiments that show that our approach substantially improves both customer retention and engagement.

Future work that we consider worth exploring is introducing impressions to AFRA to improve temporal diversity by organically down-ranking items the user has already seen but not interacted with. Furthermore, we would like to abstract the prediction head of our recommender system to allow the flexibility of re-ranking and hence the ability to apply AFRA on use cases with many millions of items.

References

1. Bracher C, Heinz S, Vollgraf R (2016) Fashion DNA: merging content and sales data for recommendation and article mapping. CoRR abs/1609.02489, http://arxiv.org/abs/1609.02489
2. Cardoso A, Daolio F, Vargas S (2018) Product characterisation towards personalisation: learning attributes from unstructured data to recommend fashion products. In: Proceedings of the 24th ACM SIGKDD international conference on knowledge discovery and data mining, KDD'18. Association for Computing Machinery, New York, NY, USA, pp 80–89, https://doi.org/10.1145/3219819.3219888
3. Celikik M, Kirmse M, Denk T, Gagliardi P, Mbarek S, Pham D, Ramallo AP (2021) Outfit generation and recommendation—An experimental study. In: Dokoohaki N, Jaradat S, Corona Pampín HJ, Shirvany R (eds) Recommender systems in fashion and retail. Springer, Cham, pp 117–137

4. Chakraborty S, Hoque MS, Rahman Jeem N, Biswas MC, Bardhan D, Lobaton E (2021) Fashion recommendation systems, models and methods: a review. Informatics 8(3). https://doi.org/10.3390/informatics8030049
5. Chen Q, Zhao H, Li W, Huang P, Ou W (2019) Behavior sequence transformer for e-commerce recommendation in Alibaba. In: Proceedings of the 1st international workshop on deep learning practice for high-dimensional sparse data, DLP-KDD'19. Association for Computing Machinery, New York, NY, USA
6. Covington P, Adams J, Sargin E (2016) Deep neural networks for youtube recommendations. In: Proceedings of the 10th ACM conference on recommender systems, RecSys'16. Association for Computing Machinery, New York, NY, USA, pp 191–198. https://doi.org/10.1145/2959100.2959190
7. Deldjoo Y, Nazary F, Ramisa A, J McAuley J, Pellegrini G, Bellogín A, Di Noia T (2022) A review of modern fashion recommender systems. arXiv:2202.02757
8. Deng Q, Wang R, Gong Z, Zheng G, Su Z (2018) Research and implementation of personalized clothing recommendation algorithm, pp 219–223. https://doi.org/10.1109/ICDH.2018.00046
9. Denk TI, Ramallo AP (2020) Contextual BERT: Conditioning the language model using a global state. arXiv:2010.15778
10. Fang H, Zhang D, Shu Y, Guo G (2019) Deep learning for sequential recommendation: algorithms, influential factors, and evaluations. https://doi.org/10.48550/ARXIV.1905.01997
11. Hidasi B, Karatzoglou A, Baltrunas L, Tikk D (2015) Session-based recommendations with recurrent neural networks. http://arxiv.org/abs/1511.06939, cite arxiv:1511.06939 Comment: Camera ready version (17th February, 2016) Affiliation update (29th March, 2016)
12. Kang W, McAuley JJ (2018) Self-attentive sequential recommendation. CoRR abs/1808.09781, http://arxiv.org/abs/1808.09781
13. Lasserre J, Sheikh AS, Koriagin E, Bergmann U, Vollgraf R, Shirvany R (2020) Meta-learning for size and fit recommendation in fashion. In: SIAM international conference on data mining (SDM20)
14. Lin Y, Moosaei M, Yang H (2019) Learning personal tastes in choosing fashion outfits. In: 2019 IEEE/CVF conference on computer vision and pattern recognition workshops (CVPRW), pp 313–315, https://doi.org/10.1109/CVPRW.2019.00041
15. Lv F, Jin T, Yu C, Sun F, Lin Q, Yang K, Ng W (2019) SDM: sequential deep matching model for online large-scale recommender system. In: Proceedings of the 28th ACM international conference on information and knowledge management, CIKM'19. Association for Computing Machinery, New York, NY, USA, pp 2635–2643. https://doi.org/10.1145/3357384.3357818
16. Mikolov T, Sutskever I, Chen K, Corrado G, Dean J (2013) Distributed representations of words and phrases and their compositionality. In: Proceedings of the 26th international conference on neural information processing systems, NIPS'13, vol 2. Curran Associates Inc., Red Hook, NY, USA, pp 3111–3119
17. Nogueira P, Gonçalves D, Queiroz Marinho V, Magalhães AR, Sá J (2021) A critical analysis of offline evaluation decisions against online results: a real-time recommendations case study
18. Pasumarthi RK, Bruch S, Wang X, Li C, Bendersky M, Najork M, Pfeifer J, Golbandi N, Anil R, Wolf S (2019) Tf-ranking: scalable tensorflow library for learning-to-rank, pp 2970–2978
19. Quadrana M, Karatzoglou A, Hidasi B, Cremonesi P (2017) Personalizing session-based recommendations with hierarchical recurrent neural networks. In: Proceedings of the eleventh ACM conference on recommender systems, RecSys'17. Association for Computing Machinery, New York, NY, USA
20. Radford A, Narasimhan K, Salimans T, Sutskever I (2018) Improving language understanding by generative pre-training (2018)
21. Rendle S, Freudenthaler C, Gantner Z, Schmidt-Thieme L (2009) BPR: Bayesian personalized ranking from implicit feedback. In: Proceedings of the twenty-fifth conference on uncertainty in artificial intelligence, UAI'09. AUAI Press, Arlington, Virginia, USA, pp 452–461
22. Sembium V, Rastogi R, Tekumalla L, Saroop A (2018) Bayesian models for product size recommendations. In: Proceedings of the 2018 world wide web conference, WWW'18, pp 679–687

23. Souza de Pereira Moreira G, Rabhi S, Lee JM, Ak R, Oldridge E, (2021) Transformers4Rec: bridging the gap between NLP and sequential/session-based recommendation. Association for Computing Machinery, New York, NY, USA

24. Steck H, Baltrunas L, Elahi E, Liang D, Raimond Y, Basilico J (2021) Deep learning for recommender systems: A netflix case study. AI Magazine 42

25. Sun F, Liu J, Wu J, Pei C, Lin X, Ou W, Jiang P (2019) Bert4rec: Sequential recommendation with bidirectional encoder representations from transformer, CIKM'19. Association for Computing Machinery, New York, NY, USA

26. Szpektor I, Maarek Y, Pelleg D (2013) When relevance is not enough: promoting diversity and freshness in personalized question recommendation. In: Proceedings of the 22nd international conference on world wide web, WWW'13. Association for Computing Machinery, New York, NY, USA, pp 1249–1260. https://doi.org/10.1145/2488388.2488497

27. Tamm YM, Damdinov R, Vasilev A (2021) Quality metrics in recommender systems: do we calculate metrics consistently? Association for Computing Machinery, New York, NY, USA, pp 708–713. https://doi.org/10.1145/3460231.3478848

28. Vaswani A, Shazeer N, Parmar N, Uszkoreit J, Jones L, Gomez AN, Kaiser L, Polosukhin I (2017) Attention is all you need. In: Proceedings of the 31st international conference on neural information processing systems, NIPS'17. Curran Associates Inc., Red Hook, NY, USA, pp 6000-6010

29. Villatel K, Smirnova E, Mary J, Preux P (2018) Recurrent neural networks for long and short-term sequential recommendation. https://doi.org/10.48550/ARXIV.1807.09142

30. Wu L, Li S, Hsieh CJ, Sharpnack J (2020) SSE-PT: sequential recommendation via personalized transformer. Association for Computing Machinery, New York, NY, USA, pp 328–337. https://doi.org/10.1145/3383313.3412258

31. Xu C, Zhao P, Liu Y, Sheng VS, Xu J, Zhuang F, Fang J, Zhou X (2019) Graph contextualized self-attention network for session-based recommendation. In: Proceedings of the 28th International joint conference on artificial intelligence, IJCAI'19. AAAI Press, pp 3940–3946

32. Ying H, Zhuang F, Zhang F, Liu Y, Xu G, Xie X, Xiong H, Wu J (2018) Sequential recommender system based on hierarchical attention network. In: Proceedings of the 27th international joint conference on artificial intelligence IJCAI'18. AAAI Press, pp 3926-3932

33. Zeng Z, Xiao C, Yao Y, Xie R, Liu Z, Lin F, Lin L, Sun M (2021) Knowledge transfer via pre-training for recommendation: a review and prospect. Front Big Data 4:602071. https://doi.org/10.3389/fdata.2021.602071

34. Zhang F, Peng Q, Wu Y, Pan Z, Zeng R, Lin D, Qi Y (2022) Multi-graph based multi-scenario recommendation in large-scale online video services. In: Companion proceedings of the web conference 2022, WWW'22, Association for Computing Machinery, New York, NY, USA, pp 1167–1175, https://doi.org/10.1145/3487553.3524729

35. Zhou Z, Di X, Zhou W, Zhang L (2018) Fashion sensitive clothing recommendation using hierarchical collocation model. In: Proceedings of the 26th ACM international conference on multimedia, MM'18. Association for Computing Machinery, New York, NY, USA, pp 1119–1127, https://doi.org/10.1145/3240508.3240596

Adversarial Attacks Against Visually Aware Fashion Outfit Recommender Systems

Matteo Attimonelli, Gianluca Amatulli, Leonardo Di Gioia, Daniele Malitesta, Yashar Deldjoo, and Tommaso Di Noia

Abstract Pre-trained CNN models are frequently employed for a variety of machine learning tasks, including visual recognition and recommendation. We are interested in examining the application of attacks generated by adversarial machine learning techniques to the vertical domain of fashion and retail products. Specifically, the present work focuses on the robustness of cutting-edge CNN models against state-of-the-art adversarial machine learning attacks that have shown promising performance in general visual classification tasks. In order to achieve this objective, we conducted adversarial experiments on two prominent fashion-related tasks: visual **clothing classification** and **outfit recommendation**. Large-scale experimental validation of the *fashion category classification* task on a real dataset of PolyVore consisting of various outfits reveals that **ResNet50** is one of the most resilient networks for the fashion categorization task, whereas *DenseNet169* and *MobileNetV2* are the most vulnerable. Performance-wise however, DenseNet169 is the most time-consuming network to attack. However, the results of the *outfit recommendation* task were somewhat unexpected. In both of the push or nuke attack scenarios and altogether, it was demonstrated that adversarial attacks were unable to degrade the quality of outfit recommenders. The only exception was the more complicated adversarial attack of **DeepFool**, which could only weaken the quality of visual recommenders at large attack budget (ϵ) values. Numerous explanations could be provided for this phenomenon, which can be attributed to the fact that a collection of adversarially perturbed images can nonetheless appear pleasing to the human eye. This may possibly be a result of the greater image sizes in the selected dataset. Overall, the results of this study are intriguing and encourage more studies in the field of adversarial attacks and fashion recommendation system security.

Keywords Adversarial · Attack · Fashion · Recommender systems

M. Attimonelli · G. Amatulli · L. D. Gioia · D. Malitesta · Y. Deldjoo (✉) · T. D. Noia
Polytechnic University of Bari, Bari, Italy
e-mail: deldjooy@acm.org

© The Author(s), under exclusive license to Springer Nature Switzerland AG 2023
H. J. Corona Pampín and R. Shirvany (eds.), *Recommender Systems in Fashion and Retail*,
Lecture Notes in Electrical Engineering 981,
https://doi.org/10.1007/978-3-031-22192-7_4

1   Introduction and Related Work

From birth until death, people are surrounded by fashion at every stage of their lives. We can make some educated guesses about a person's personality, lifestyle, and social-working class just based on how they present themselves to the world through their clothing. During the last few years, the textile and clothing industries have evolved dramatically. Millions of products are now available in online catalogs, so customers no longer have to visit multiple stores, wait in long lines, or try on outfits in dressing rooms. However, in light of the abundance of choices, an effective recommendation system (RS) is needed to sort, order, and present product information to users. Deldjoo et al. [11] provide an overview of the literature on contemporary fashion recommender systems. In their work, the authors develop a taxonomy for fashion RS research that classifies the literature according to the goals these systems seek to achieve, namely *item recommendation*, *outfit recommendation*, and *size recommendation*, as well as the type of side-information they employ, namely side-information of users, items, and context. Cheng et al. [6] study application of computer vision techniques approaches on fashion products, defining four areas that characterize current research development in the field: fashion detection, fashion analysis, fashion synthesis, and fashion recommendation. As stated in [11], one of the challenging goals in the textile sector is to coordinate a variety of fashionable items in such a way that they, when worn together as an outfit, provide a pleasing visual experience. Measuring the quality of an **outfit** is crucial in order to devise an effective fashion recommendation system.

Training a model that can determine whether or not two or several items go together, often known as compatibility, is intrinsically a challenging task. Recent research has examined various item compatibility signals, such as co-purchase data [20, 30], outfits created by professional fashion designers [15], and outfits discovered by studying what people wear in social media photographs [29]. From this compatibility information, associated picture and text data are then used to understand the generalization of compatible products. Much of the success in the visual compatibility prediction tasks owes to rise of deep learning, in particular, convolutional neural networks (CNNs), or driven models such as Siamese-CNN, consider [9, 24, 26, 32, 33] for a few prominent examples (Fig. 1).

Notwithstanding their remarkable success, recent development in adversarial machine learning research has highlighted the vulnerability of (neural) recommender systems to **adversarial attacks** [4, 10], especially when considering the task of single-item recommendation [1]. Attacks on visually aware RS such as in the fashion domain are more effective when adversarial images are produced that are similar to the source images (pixel-level perturbations) and can degrade the quality of the RSs. In the fashion domain, adversarial attacks have been used against images classification, for example, for pushing (promoting) or nuking (demoting) specific categories of products [2, 23]. Anelli et al. [3], for instance, describe how certain adversarial attacks against CNNs strategies may negatively impact the popularity of items. These modest adversarial perturbations could cause deep learning algorithms to misclassify incoming data. Due to the importance of overcoming this issue, Xu et al. [31]

Fig. 1 Example of a "Good Outfit" recommended correctly by our outfit recommender model (before attack)

outline numerous defenses against adversarial examples, categorizing them into three distinct groups: gradient masking/obfuscation, robust optimization, and adversarial examples detection.

Despite recent progress, the majority of this research has focused on attacks against fashion classification systems, such as fashion item category classification and single-item recommendation. Rather than focusing solely on individual item recommendations, in this research we aim to alter the overall classification of an outfit grader, which we deem to be more important due to its bigger impact on revenue. The transferability [25] of adversarial instances generated by CNNs is an important problem. Transferability in this sense refers to the fact that adversarial occurrences designed to deceive one CNN are commonly misclassified by other CNNs. We are particularly interested in exploring this subject for fashion item classification and outfit recommendation tasks.

The main contribution of our work is the following:

- **Adversarial attack against <u>two</u> prominent fashion prediction tasks**. We have investigated the effects of four state-of-the-art adversarial attack models on a suite of four widely adopted pretrained CNN architectures in the field of computer vision and visually aware recommendation [9], with a focus on the fashion **category classification** and **outfit recommender** (grader) systems.
- **Evaluation**. One key distinction of our work is that it relies on PolyVore [15] which is a dataset that reflects the actual world, as opposed to popular idealized datasets such as MNIST and fashion-MNIST. The dataset is a crowdsourced dataset created by the users of a website that permitted its members to submit images of fashion products and categorize them into outfits. We used PolyVore for training, launching attacks, and evaluating performances; in particular, we created two distinct datasets for the classification system and the recommendation task using PolyVore's data.

This paper is organized as follows: In Sect. 2, we introduce technical details regarding the CNN and attacks explored during our study; in Sect. 3, we discuss the setup employed to build and test the fashion classification and outfit recommendation frameworks; in Sect. 4, we investigate about the impact of the attacks on our models and discuss the results; finally in Sect. 5, we formulate conclusions.

2   Visual Attacks Against Fashion Classifiers

Several image classifier architectures have been presented throughout the years with the goal of achieving the maximum accuracy possible on the test set. Recent architectures have achieved this goal in part; our purpose is to determine what happens when we target these classifiers via adversarial attacks in the context of fashion category classification and how much their performance degrades.

Problem formulation. Consider a dataset $\mathcal{D}$ composed by n pairs $\{(x, y)\}_{i=1}^{i=n}$, where x is the input sample and y is the corresponding label, we can formalize the classification task as finding a mathematical function $f_\theta : \mathcal{X} \to \mathcal{Y}$ such that—for each input x—it could predict y, exploiting the knowledge acquired in the parameters' matrix θ. The objective of the adversarial attacks is to find a perturbation δ, following a specific strategy, to produce an adversarial example $x^* = x + \delta$ such that $f(x^*) \neq y$, so which causes mis-classification. Note that, since fashion outfit recommendation is essentially a classification system formed as outfit grading, there is no need to design a new mis-recommendation task here. The design of δ is the objective of the *adversarial attack* methods. The research of adversarial assaults is motivated by a number of factors [5]: Firstly, it can assist in identifying models' vulnerabilities before they are exploited, and secondly, it enables the investigation of the impact of each attack and the subsequent deployment of countermeasures.

2.1   *Explored Adversarial Attacks*

Numerous algorithms for adversarial attacks have been proposed over the years. To evaluate the resistance of our fashion ML model against adversarial attacks, we have studied the following state-of-the-art attack models:

- **FGM** [14]: It is a generalization of the FGSM attack that uses the L2 norm of the gradient of the cost function instead of its sign. The objective is to find a minimum perturbation $\mathbf{x}^*$ that maximizes the cost function employed for the training $J(\mathbf{x}^*, y)$—aiming to worsening model performances—maintaining the L2 norm bounded, such that $||x^* - x||_2 < \epsilon$, where ϵ is the attack strength an ∇_x represents the gradient of the cost function. The attack can be formalized as follows:

$$\mathbf{x}^* = \mathbf{x} + \epsilon \cdot \frac{\nabla_x J(\mathbf{x}, y)}{||\nabla_x J(\mathbf{x}, y)||_2}$$

- **BIM** [18]: It is an iterative version of FGM and FGSM, summarizable as it follows:

$$\mathbf{x}_0^* = \mathbf{x},$$

$$\mathbf{x}_{N+1}^* = \mathbf{x}_N^* + \epsilon \cdot \frac{\nabla_x J(\mathbf{x}, y)}{||\nabla_x J(\mathbf{x}, y)||_2}$$

where $\mathbf{x}_0^*$ is the original test sample, $\mathbf{x}_{N+1}^*$ is the N+1-th step of the method, and ϵ is the attack's budget.

- **DeepFool** [21]: It is one of the most recent attack algorithms whose goal is to provide a method for calculating adversarial perturbations that can also be used as adversarial training samples to enhance the robustness of classifiers. As inputs, the algorithm receives samples and a classifier and adds a little perturbation at each step until the projected label is erroneous.
- **Projected Gradient Descent (PGD)** [1, 19]: This attack involves the repeated projection of a disturbance onto an l_p-ball. Re-projections occur at the conclusion of each loop. PGD initializes the example at a random location on the ball of interest and executes random restarts, whereas BIM initializes to the starting position.

2.2 CNN Architectures

In our work, we have considered four state-of-the-art models:

- **VGG16** [28]: It is a very complex/deep neural network design, with 13 convolutional layers and three dense ones. Smaller filter sizes in VGG16 (particularly 3×3 for convolutional layers and 2×2 for max-pooling ones) enable for more significant features to be captured in the input patterns. As a result, this network type requires huge memory for processing.
- **ResNet50** [16]: It was the winner of the ImageNet challenge in 2015, and it consists of 50 layers, where the introduction of residual connections has been its most significant contribution. These connections not only make it possible to solve the problem of shattered gradient, but they also allow to keep information regarding the input patterns, even when the forwarding phase is in progress.
- **DenseNet169** [17]: There are fewer parameters required. Its primary contribution is the connection between each layer and the others, meaning that all precomputed feature maps serve as input for the subsequent block. This improves the propagation of features and minimizes the number of parameters.

- **MobileNetV2** [27]: It is built on an inverted residual structure, with shortcut connections between bottleneck levels. This architecture is highly adaptable and can be used to tackle a variety of problems, such as the foundation for object identification and semantic segmentation models.

3 Experimental Setup

In this section, we describe how we have employed each model and how we have performed the attacks. Also, we talk about the dataset that we have designed—which allows to train the models and perform the attacks.

3.1 Dataset

The dataset employed in this work is derived from PolyVore [15], a real-world dataset related to fashion outfits. It includes 21889 outfits split in 17316 outfits for training, 1497 outfits for validation, and 3076 for test.

In order to perform fashion classification, we have modified the original dataset creating a new one that includes nine fashion categories: "boot", "coat", "dress", "jeans", "shoe", "short", "skirt", "top", and "t-shirt". This is possible due to the fact that the original dataset consists of separate json files—one for each of the training, validation, and test sets—that record information about each outfit, its constituent pieces, and its category. In conclusion, our dataset consists of 17334, 1195, and 2401 training, validation, and test samples, each measuring 224×224 pixels $\times$ pixels.

The recommendation dataset is instead derived from PolyVore's fashion compatibility dataset, which has over 7000 outfits categorized as "compatible" and "incompatible". From this, we built training and test sets consisting of 6076 and 1000 samples, respectively.

3.2 ML Model Implementation

Classification model. We opted to employ transfer learning on pretrained versions of the ImageNet dataset [13] to study the resilience of cutting-edge image classification systems against state-of-the-art adversarial attacks for the fashion categorization task.

To adapt the models to our dataset, we leveraged their pretrained backbones and added our own customized heads. In order to replicate the performances of the original classifiers, the new derived models are intended to fit nine classes—as opposed to the 1000 in ImageNet. With the exception of the final dense layer, which consists of nine units, each head has the same structure as the original. After running a series of experiments, we found that defrosting the backbone weights and training with a

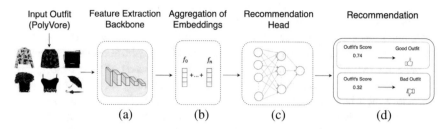

Fig. 2 Overall fashion recommender system adopted in our framework. In the first stage (**a**), a pretrained CNN backbone extracts latent features for each fashion item image. After an aggregation step (**b**) involving the element-wise summation of features, a neural architecture (**c**) is responsible for the recommendation generation, which calculates the degree of compatibility to have a matching outfit (**d**)

very low learning rate (precisely 0.0001) using the Adam optimizer was the most effective strategy for applying transfer learning to our models in an effort to improve their performance. Fine-tuning has been carried out over 20 epochs.

Outfit recommendation model. The basis of the outfit recommendation model is a fashion classification model, which has been modified for the outfit recommendation problem. The goal of the recommender system is to grade an individual's outfit based on whether or not it contains "Good" or "Bad" combinations of dresses. We have separated the architecture into its two primary components in order to construct the recommendation model: (i) a core feature extraction mechanism and (ii) a central recommendation processing unit (Fig. 2).

The first step is responsible for extracting features from each image that belongs to an outfit, followed by the addition of all the embeddings of the same outfit to create a single latent representation for each outfit. ResNet50 has been utilized as the network's backbone for the reasons described in Sect. 4. As recommendation head, we have instead chosen a fully connected neural network composed by one DropOut unit—in order to prevent overfitting—placed between two dense layers, the first of which has 2048 units—as the number of elements stored in each latent representation—and the second of which contains only one unit. The recommendation head receives as input the embeddings associated with each outfit and outputs the likelihood that the outfit's components are combined in a proper way or not, hence rating the outfit as "Good" or "Bad". In contrast to the feature extraction backbone, which was already pretrained on the ImageNet dataset, the recommendation head was trained for 100 epochs using the Adam optimizer.

3.3 Attacks Implementation

The evasion attacks were conducted using the Adversarial Robustness Toolbox (ART) from IBM [22]. This toolkit enables the execution of a variety of cutting-edge attacks on particular datasets and models. It is possible to configure a number of attack-specific and non-specific variables, such as the attack's budget and batch size, for each attack. Regarding the recommender system's attacks, we utilized all of those mentioned earlier in Sect. 2.1 with the exception of PGD, whose results are identical to those of BIM. After trying many combinations, we noticed that the attacks against outfit recommendation are powerful at extremely high budgets, thus we adopt an extremely high budget—specifically 50. As a maximum number of iterations for iterative approaches, we have selected a maximum of 10. The goal of the attacks is to modify the results obtained by the recommender by attacking the CNN, which is responsible for extracting all of the properties of the clothing that are utilized by the recommendation head.

4   Results and Discussion

Classification Results. In Table 1, we present all the findings achieved in terms of accuracy before (**Base**) and after attacks (by attacking models **FGM**, **BIM**, **Deep-Fool**, **PGD**), the attacks' efficacy denoted with ratio (normalized attacks' values against **Base**), and the time required to execute all attacks for each budget value. Looking at the data pertaining to the initial condition (marked with **Base** related to before executing any attacks), we can conclude that DenseNet169 has the best accuracy with 85.92% and ResNet50 has the lowest accuracy with 83.72%. As it can be seen, however, each of the four figures falls inside a relatively narrow band, between 83 and 86% accuracy. We also opted to provide the average value of all these variables so that the model analysis could be conducted more easily. In reality, adjusting the study to the average accuracy following attacks with a budget of $\epsilon = 0.02$, we can note that DenseNet169 has the lowest value (16.01%) compared to ResNet50, which has the highest value (21.41%). The range of these numbers is from 16 to 22%, indicating that no model is particularly resistant to attack. By evaluating the whole data for the same budget value selected in advance, we could make the following observations:

- From an **attacker's point of view**: **DeepFool** is the most aggressive attack in terms of post-attack accuracy values. For instance, looking at the attack success ratios we can see that on **ResNet50** the ratios for DeepFool versus FGM are **0.16** versus **0.40**, while for **VGG16** they are **0.14** versus **0.36**. As for the CNN type, VGG16 is the one that requires the least amount of time to execute the task, relative to the other models included in this study. Also, **ResNet50** appears to be the most resilient model, as its average accuracy after an attack is approximately 21% compared to 16% and 17% shown by DenseNet169 and MobileNetV2, respectively. On the other hand, *DenseNet169* appears to be the least resistant network, with an average accuracy after attack value of 16%, making it an obvious candidate for various attacks.
- From the **architecture robustness point of view**: ResNet50 appears to be the strongest network followed by VGG16, MobileNetV2, and DenseNet169. Addi-

Table 1 Conducted adversarial attacks on CNN's state-of-the-art fashion category classification task

| | Performed adversarial attacks | | | |
| | Adv attack | $\epsilon = 0.02$ | | |
		Accuracy (%)	Time (s)	Ratio
ResNet50	Base	83.72	/	/
	FGM	33.61	71.94	0.40
	BIM	19.03	3191.06	0.23
	DeepFool	13.16	9004.18	0.16
	PGD	19.82	3384.91	0.24
	Avg	**21.41**	**3913.02**	**0.26**
VGG16	Base	84.92	/	/
	FGM	31.32	25.92	0.36
	BIM	17.62	931.18	0.21
	DeepFool	12.20	1177.77	0.14
	PGD	17.61	992.65	0.21
	Avg	**19.69**	**781.88s**	**0.23**
MobileNetV2	Base	85.42	/	/
	FGM	24.98	61.72	0.29
	BIM	15.32	2791.17	0.18
	DeepFool	12.32	3934.81	0.14
	PGD	15.24	3128.25	0.18
	Avg	**16.97**	**2478.99**	**0.20**
DenseNet169	Base	85.92	/	/
	FGM	22.57	158.07	0.26
	BIM	14.45	7423.79	0.17
	DeepFool	12.41	29149.52	0.14
	PGD	14.62	7487.82	0.17
	Avg	**16.01**	**11054.80**	**0.19**

The base model displays the ML classifier prior to an attack. The ratio reflects the effectiveness of the attacks

tionally, ResNet50 requires the second highest amount of time—among the CNN tested—to be attacked, and it is second only to DenseNet169. The adoption of ResNet50 as the defensive strategy may therefore be wise, given that it is both the most resilient (greatest accuracy after assault) and one of the most time-consuming to attack.

BIM and PGD have nearly same required time and accuracy values, hence selecting one over the other is an equally weighted decision. As stated before, we also conducted all attacks with three distinct budgets as given in Table 2 and chose to explore this variance with a focus on ResNet50 and VGG16. Unlike the prior stated characteristics, it appears that the average accuracy of an attack decreases as epsilon grows. With $\epsilon = 0.05$, VGG16 looks to be less resilient than ResNet50, but the opposite is

Table 2 Conducted adversarial attacks on CNN's state-of-the-art fashion classification system, with budget variation

| | | Performed adversarial attacks | | | | | | | | |
| | | $\epsilon=0.01$ | | | $\epsilon=0.02$ | | | $\epsilon=0.05$ | | |
	Adv attack	Acc. (%)	Time (s)	Ratio	Acc. (%)	Time (s)	Ratio	Acc. (%)	Time (s)	Ratio
ResNet50	Base	83.72	/	/	83.72	/	/	83.72	/	/
	FGM	36.19	66.27	0.43	33.61	71.94	0.40	36.07	67	0.43
	BIM	27.57	3001.55	0.33	19.03	3191.06	0.23	14.20	2998	0.17
	DeepFool	13.04	7853.76	0.16	13.16	9004.18	0.16	15.73	3158	0.19
	PGD	27.57	3268.72	0.33	19.82	3384.91	0.24	14.29	3308.45	0.17
	Avg	**26.09**	**3547.58**	**0.32**	**21.41**	**3913.02**	**0.26**	**20.07**	**2382.86**	**0.24**
VGG16	Base	84.92	/	/	84.92	/	/	84.92	/	/
	FGM	40.77	27.46	0.48	31.32	25.92	0.36	24.60	24.72	0.29
	BIM	30.15	993.74	0.36	17.62	931.18	0.21	7.50	1015	0.09
	DeepFool	12.12	1181.44	0.14	12.20	1177.77	0.14	12.50	1158	0.15
	PGD	30.19	996.98	0.36	17.61	992.65	0.21	7.37	1733.57	0.09
	Avg	**28.31**	**799.91**	**0.34**	**19.69**	**781.88**	**0.23**	**13.00**	**982.82**	**0.16**

true when $\epsilon = 0.01$ is used. As epsilon increases, the effects of assaults become more visible to the human sight, which is an additional factor that must be considered. This increases the potency of these attacks while emphasizing the need for more caution.

Recommendation results. This study compares the recommendations produced from the execution of three distinct forms of adversarial attacks with three different budget levels. Through the execution of previously described adversarial attacks, namely FGM, BIM, and DeepFool, we assessed them in terms of Loss for "Yes" and "No" probability distributions. We define as "Yes" the probability distribution associated with "Good Outfits" and as "No" the distribution associated with "Bad Outfits". In this form of classification, we may identify a "Good Outfit" if its value falls between [0.51] and [1] and a "Bad Outfit" if its value falls between [0] and [0.50]. We refer to No→Yes label-flipping as a **"push attack"** and a **"nuke attack"** for Yes→No. According to the results given in Table 3, **DeepFool>BIM>FGM** (strongest to weakest) in terms of overall accuracy; however, note that this is for the strongest attack budget of $\epsilon = 50$. Due to computational and storage limitations, DeepFool has been implemented with a single budget value of 50 for its epsilon parameter. When studying the Push or Nuke attack (on the Yes and No class), we see that *BIM* and *FGM* are ineffective compared to the base; for example, even at a high budget of $\epsilon = 50$, the accuracies for the base are approximately 54.60% (Yes) and 88.59% (No), respectively. Note that despite the fact that BIM is more powerful than FGM for the push attack, none of them can deteriorate the quality below the base (54.6%). Only **DeepFool** can slightly degrade the base accuracy on push attack (88.59% → 78.60%), but it remains ineffective in nuke attack (54.6% → 62.19%). Therefore, the overall summary of this study is that outfit recommendation remains a substantially more difficult assignment than single-item classification. Only advanced adversarial techniques, such as **DeepFool**, are able to compromise

Table 3 Results of the conducted adversarial attacks on the **fashion outfit recommendation** system

Adversarial attack	Budget (ϵ)	Loss (good, bad)		Accuracy (good, bad) (%)		Loss (O.D.)	Acc. (O.D.) (%)
Base	/	0.7595	0.3674	54.60	88.59	0.5634	71.60
FGM	0.02	0.7593	0.3675	54.60	88.59	0.5634	71.60
BIM	0.02	0.7593	0.3675	54.60	88.59	0.5634	71.60
FGM	5	0.6761	0.4202	61.00	83.20	0.5481	72.10
BIM	5	0.7545	0.3706	54.79	88.59	0.5625	71.70
FGM	50	0.6530	0.4195	63.20	83.00	0.5362	73.10
BIM	50	0.7542	0.3706	54.79	88.59	0.5624	71.70
DeepFool	50	0.6325	0.5016	62.19	78.60	0.5671	70.39

Note that we have only employed ResNet50 as the core model due to its superior performance in the previous fashion classification task. The figures underneath the "Yes" and "No" categories indicate the success of label-flipping against positive and negative classes. It could be observed, for instance, that flipping a "No" class (No→Yes) is easier than a "Yes" class (Yes→No). *Note* "O.D." stands for "Overall Dataset" data

Fig. 3 Visualizing the impact of a randomly chosen outfit with ID 956 before (top) and after (bottom) DeepFool attack

the robustness, and even then, only at a high cost. Insightful conclusions may be drawn from this, such as the fact that it is far more difficult to **undermine** what a user considers as an aesthetically Good Outfit on a group (outfit) compared to a single image.

Qualitative example. We provide here a graphic representation of adversarial attacks we conducted against fashion outfits on the PolyVore dataset. We selected a random outfit with ID:956. Looking at the results of the obtained attacks, we can conclude that DeepFool is the only attack that demonstrates its power, with a "Before Attack Classification" value of 0.2541, indicating a "Bad Outfit," and a "After Attack Classification" value of 0.5378, indicating a "Good Outfit" (Fig. 3).

We choose to execute these attacks with a budget of 50 due to the data presented in Table 3, which indicate that by increasing the epsilon value of the attack, we can achieve slightly better results. The illustration depicts the results of a DeepFool attack on Outfit 956 from our test set. We computed the structural similarity index (SSIM) between two pairs of similar images (before and after attack) in order to visually note the effects of the attacks on the outfit images. The sequence of images at the top row shows the original items, but the sequence at the bottom depicts the attacked items; as can be seen, the results of the adversarial attack are not readily apparent in the photographs. In fact, we found the following similarity values for the four items: 0.9893 (item 1 jacket), 0.9977 (item 2 dress), 0.7491 (item 3 bag), and 0.6469 (item 4 necklace). Analyzing the collected similarity data reveals that the presence of additional noise has a significant effect on the photos. As can be seen, the values associated with item 1 (Jacket) and item 2 (Dress) are close to 1.00, which, in SSIM words, indicates that the two images before and after the attack are quite similar. In contrast, the other two products, item 3 (Bag) and item 4 (Necklace), have lower similarity levels. A possible reason could be that, upon studying the images

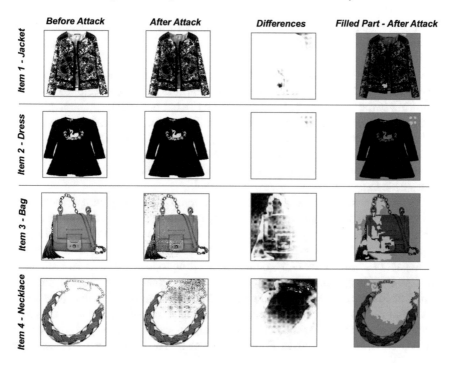

Fig. 4 SSIM comparison between each item of the selected outfit 956

associated with these objects, it is evident to the human eye that noise is present in the after attack photos. As evidenced by the analysis of Fig. 4, in order to better explain this phenomena and highlight the areas of the images that appear to have been altered, we generated Disparity Maps, which depict graphically where the alterations are located.

5 Conclusions

We proposed the use of adversarial attacks, including FGM, BIM, DeepFool, and PGD, on two popular fashion prediction tasks, namely *fashion category classification* and **outfit recommendation**. This was accomplished by executing attacks on various CNNs that had been pretrained, including ResNet50, VGG16, DenseNet169, and MobileNetV2. The dataset of the fashion images is based on PolyVore, which differs significantly from smaller-scale datasets such as MNIST and fashion-MNIST.

On *garment category classification*, the results of a large-scale experimental validation consisting of 16 different (attack, CNN) combinations show that adversarial attacks succeed best on *DenseNet169* (the weakest CNN), whereas **ResNet50** is the most secure against such attacks. DenseNet169 and ResNet50 require the most time for penetration. ResNet50 has proven to be one of the most reliable networks for the fashion classification task, while DenseNet169 and MobileNetV2 have demonstrated the lowest levels of performance.

We have also evaluated the impact of three different assaults—FGM, BIM, and DeepFool—on the fashion recommender system that we have developed, and the following two findings are noteworthy: (i) The **DeepFool** attack is the most potent of those investigated because it can both degrade the overall accuracy on the test set and cause the fashion recommender to fail by recommending poorly built ensembles as good ones; (ii) the fact that the accuracy before and after a DeepFool attack differs by only 1.21% reveals the fact that impact of such an attack on the images—even with such a high budget—is not really noticeable.

For future work, we believe it is vital to develop more complex approaches for recommending useful fashion outfits. We are also interested in conducting and analyzing attacks at a more granular level, such as by assessing the vulnerability of user-item or user-outfit classes (see a similar work in [7]). In an effort to decrease the impact of evasion attacks, it could be interesting to identify the most prominent portions of an image that help attacks' effectiveness. Finally, we discover a plethora of multimedia recommender systems applications utilizing visual content (e.g., meal and restaurant recommendation, media and news recommendation, visual query recommendation) that may be subject to adversarial attacks for a variety of commercial and non-commercial incentives. For more information on these systems, please refer to [8, 12].

References

1. Anelli VW, Bellogin A, Deldjoo Y, Di Noia T, Merra FA (2021) Msap: multi-step adversarial perturbations on recommender systems embeddings. In: The 34th international FLAIRS conference. The Florida AI Research Society (FLAIRS), AAAI Press, pp 1–6
2. Anelli VW, Deldjoo Y, Noia TD, Malitesta D, Merra FA (2021) A study of defensive methods to protect visual recommendation against adversarial manipulation of images. In: Diaz F, Shah C, Suel T, Castells P, Jones R, Sakai T (eds) SIGIR '21: the 44th international ACM SIGIR conference on research and development in information retrieval, virtual Event, Canada, 11–15 July 2021. ACM, pp 1094–1103. https://doi.org/10.1145/3404835.3462848
3. Anelli VW, Di Noia T, Di Sciascio E, Malitesta D, Merra FA (2021c) Adversarial attacks against visual recommendation: an investigation on the influence of items' popularity. In: Proceedings of the 2nd workshop on online misinformation-and harm-aware recommender systems (OHARS 2021), Amsterdam, Netherlands
4. Anelli VW, Deldjoo Y, DiNoia T, Merra FA (2022) Adversarial recommender systems: attack, defense, and advances. In: Recommender systems handbook. Springer, pp 335–379
5. Biggio B, Corona I, Maiorca D, Nelson B, Srndic N, Laskov P, Giacinto G, Roli F (2017) Evasion attacks against machine learning at test time. CoRR abs/1708.06131, http://arxiv.org/abs/1708.06131, eprint1708.06131
6. Cheng W, Song S, Chen C, Hidayati SC, Liu J (2021) Fashion meets computer vision: a survey. ACM Comput Surv 54(4):72:1–72:41. https://doi.org/10.1145/3447239
7. Deldjoo Y, Di Noia T, Merra FA (2019) Assessing the impact of a user-item collaborative attack on class of users. In: ImpactRS@RecSys'19 workshop on the impact of recommender systems

8. Deldjoo Y, Schedl M, Cremonesi P, Pasi G (2020) Recommender systems leveraging multimedia content. ACM Comput Surv (CSUR) 53(5):1–38
9. Deldjoo Y, Noia TD, Malitesta D, Merra FA (2021) A study on the relative importance of convolutional neural networks in visually-aware recommender systems. In: IEEE conference on computer vision and pattern recognition workshops, CVPR Workshops 2021, virtual, 19–25 June 2021. Computer Vision Foundation/IEEE, pp 3961–3967. https://doi.org/10.1109/CVPRW53098.2021.00445. https://openaccess.thecvf.com/content/CVPR2021W/CVFAD/html/Deldjoo_A_Study_on_the_Relative_Importance_of_Convolutional_Neural_Networks_CVPRW_2021_paper.html
10. Deldjoo Y, Noia TD, Merra FA (2021) A survey on adversarial recommender systems: from attack/defense strategies to generative adversarial networks. ACM Comput Surv 54(2):35:1–35:38. https://doi.org/10.1145/3439729
11. Deldjoo Y, Nazary F, Ramisa A, McAuley J, Pellegrini G, Bellogín A, Noia TD (2023) A review of modern fashion recommender systems. ACM Comput Surv
12. Deldjoo Y, Schedl M, Hidasi B, Wei Y, He X (2022) Multimedia recommender systems: algorithms and challenges. In: Recommender systems handbook. Springer, pp 973–1014
13. Deng J, Dong W, Socher R, Li L, Li K, Fei-Fei L (2009) Imagenet: a large-scale hierarchical image database. In: 2009 IEEE Computer Society conference on computer vision and pattern recognition (CVPR 2009), 20–25 June 2009, Miami, FL, USA. IEEE Computer Society, pp 248–255. https://doi.org/10.1109/CVPR.2009.5206848
14. Goodfellow IJ, Shlens J, Szegedy C (2015) Explaining and harnessing adversarial examples. In: Bengio Y, LeCun Y (eds) 3rd international conference on learning representations, ICLR 2015, San Diego, CA, USA, 7–9 May 2015, Conference Track Proceedings. http://arxiv.org/abs/1412.6572
15. Han X, Wu Z, Jiang Y, Davis LS (2017) Learning fashion compatibility with bidirectional LSTMs. In: Liu Q, Lienhart R, Wang H, Chen SK, Boll S, Chen YP, Friedland G, Li J, Yan S (eds) Proceedings of the 2017 ACM on multimedia conference, MM 2017, Mountain View, CA, USA, 23–27 October 2017. ACM, pp 1078–1086. https://doi.org/10.1145/3123266.3123394
16. He K, Zhang X, Ren S, Sun J (2016) Deep residual learning for image recognition. In: 2016 IEEE conference on computer vision and pattern recognition, CVPR 2016, Las Vegas, NV, USA, 27–30 June 2016. IEEE Computer Society, pp 770–778. https://doi.org/10.1109/CVPR.2016.90
17. Huang G, Liu Z, Weinberger KQ (2016) Densely connected convolutional networks. CoRR abs/1608.06993, http://arxiv.org/abs/1608.06993, eprint1608.06993
18. Kurakin A, Goodfellow IJ, Bengio S (2017) Adversarial examples in the physical world. In: 5th international conference on learning representations, ICLR 2017, Toulon, France, 24–26 April 2017, Workshop track proceedings, OpenReview.net. https://openreview.net/forum?id=HJGU3Rodl
19. Madry A, Makelov A, Schmidt L, Tsipras D, Vladu A (2018) Towards deep learning models resistant to adversarial attacks. In: 6th international conference on learning representations, ICLR 2018, Vancouver, BC, Canada, April 30–May 3, 2018, Conference Track Proceedings, OpenReview.net. https://openreview.net/forum?id=rJzIBfZAb
20. McAuley J, Targett C, Shi Q, Van Den Hengel A (2015) Image-based recommendations on styles and substitutes. In: Proceedings of the 38th international ACM SIGIR conference on research and development in information retrieval, pp 43–52
21. Moosavi-Dezfooli S, Fawzi A, Frossard P (2016) Deepfool: a simple and accurate method to fool deep neural networks. In: 2016 IEEE conference on computer vision and pattern recognition, CVPR 2016, Las Vegas, NV, USA, 27–30 June 2016. IEEE Computer Society, pp 2574–2582. https://doi.org/10.1109/CVPR.2016.282
22. Nicolae MI, Sinn M, Tran MN, Buesser B, Rawat A, Wistuba M, Zantedeschi V, Baracaldo N, Chen B, Ludwig H et al (2018) Adversarial robustness toolbox v1. 0.0. arXiv preprint arXiv:1807.01069
23. Noia TD, Malitesta D, Merra FA (2020) TAaMR: targeted adversarial attack against multimedia recommender systems. In: DSN workshops. IEEE, pp 1–8

24. Pan T, Dai Y, Tsai W, Hu M (2017) Deep model style: cross-class style compatibility for 3D furniture within a scene. In: Nie J, Obradovic Z, Suzumura T, Ghosh R, Nambiar R, Wang C, Zang H, Baeza-Yates R, Hu X, Kepner J, Cuzzocrea A, Tang J, Toyoda M (eds) 2017 IEEE international conference on big bata (IEEE BigData 2017), Boston, MA, USA, 11–14 Dec 2017. IEEE Computer Society, pp 4307–4313. https://doi.org/10.1109/BigData.2017.8258459
25. Pillai RS, Sreekumar K (2020) Classification of fashion images using transfer learning. In: Bhateja V, Peng S, Satapathy SC, Zhang Y (eds) Evolution in computational intelligence—frontiers in intelligent computing: theory and applications (FICTA 2020), vol 1, Karnataka, Surathkal, India, 4–5 Jan 2020. Advances in intelligent systems and computing, vol 1176. Springer, pp 325–332. https://doi.org/10.1007/978-981-15-5788-0_32
26. Polanía LF, Gupte S (2019) Learning fashion compatibility across apparel categories for outfit recommendation. In: 2019 IEEE international conference on image processing, ICIP 2019, Taipei, Taiwan, 22–25 Sept 2019. IEEE, pp 4489–4493. https://doi.org/10.1109/ICIP.2019. 8803587
27. Sandler M, Howard AG, Zhu M, Zhmoginov A, Chen L (2018) Mobilenetv2: inverted residuals and linear bottlenecks. In: 2018 IEEE conference on computer vision and pattern recognition, CVPR 2018, Salt Lake City, UT, USA, 18–22 June 2018. Computer Vision Foundation/IEEE Computer Society, pp 4510–4520. https://doi.org/10.1109/CVPR. 2018.00474. http://openaccess.thecvf.com/content_cvpr_2018/html/Sandler_MobileNetV2_ Inverted_Residuals_CVPR_2018_paper.html
28. Simonyan K, Zisserman A (2015) Very deep convolutional networks for large-scale image recognition. In: Bengio Y, LeCun Y (eds) 3rd international conference on learning representations, ICLR 2015, San Diego, CA, USA, 7–9 May 2015. Conference track proceedings. http:// arxiv.org/abs/1409.1556
29. Sun G, Cheng Z, Wu X, Peng Q (2018) Personalized clothing recommendation combining user social circle and fashion style consistency. Multim Tools Appl 77(14):17731–17754
30. Veit A, Kovacs B, Bell S, McAuley J, Bala K, Belongie SJ (2015) Learning visual clothing style with heterogeneous dyadic co-occurrences. In: 2015 IEEE international conference on computer vision, ICCV 2015, Santiago, Chile, 7–13 Dec 2015. IEEE Computer Society, pp 4642–4650. https://doi.org/10.1109/ICCV.2015.527
31. Xu H, Ma Y, Liu H, Deb D, Liu H, Tang J, Jain AK (2020) Adversarial attacks and defenses in images, graphs and text: a review. Int J Autom Comput 17(2):151–178
32. Yin R, Li K, Lu J, Zhang G (2019) Enhancing fashion recommendation with visual compatibility relationship. In: Liu L, White RW, Mantrach A, Silvestri F, McAuley J, Baeza-Yates R, Zia L (eds) The world wide web conference, WWW 2019, San Francisco, CA, USA, 13–17 May 2019. ACM, pp 3434–3440. https://doi.org/10.1145/3308558.3313739
33. Zhao K, Hu X, Bu J, Wang C (2017) Deep style match for complementary recommendation WS-17. http://aaai.org/ocs/index.php/WS/AAAIW17/paper/view/15069

Contrastive Learning
for Topic-Dependent Image Ranking

Jihyeong Ko, Jisu Jeong, and Kyumgmin Kim

Abstract In e-commerce, users' feedback may vary depending on how the information they encounter is structured. Recently, ranking approaches based on deep learning successfully provided good content to users. In this line of work, we propose a novel method for *selecting the best from multiple images considering a topic*. For a given product, we can commonly imagine selecting the representative from several images describing the product to sell it with intuitive visual information. In this case, we should consider two factors: (1) *how attractive each image is to users* and (2) *how well each image fits the given product concept (i.e., topic)*. Even though it seems that existing ranking approaches can solve the problem, we experimentally observed that they do not consider the factor (2) correctly. In this paper, we propose CLIK (Contrastive Learning for topic-dependent Image ranKing) that effectively solves the problem by considering both factors simultaneously. Our model performs two novel training tasks. At first, in *topic matching*, our model learns the semantic relationship between various images and topics based on contrastive learning. Secondly, in *image ranking*, our model ranks given images considering a given topic leveraging knowledge learned from *topic matching* using contrastive loss. Both training tasks are done simultaneously by integrated modules with shared weights. Our method showed significant offline evaluation results and had more positive feedback from users in online A/B testing compared to existing methods.

Jihyeong Ko: Work done while intern at NAVER CLOVA

J. Ko (✉)
WATCHA Inc., NAVER CLOVA, Seoul, South Korea
e-mail: louis.ko@watcha.com

J. Jeong · K. Kim
NAVER CLOVA, NAVER AI LAB, Seongnam, South Korea
e-mail: jisu.jeong@navercorp.com

K. Kim
e-mail: kyungmin.kim.ml@navercorp.com

© The Author(s), under exclusive license to Springer Nature Switzerland AG 2023   79
H. J. Corona Pampín and R. Shirvany (eds.), *Recommender Systems in Fashion and Retail*,
Lecture Notes in Electrical Engineering 981,
https://doi.org/10.1007/978-3-031-22192-7_5

1 Introduction

In e-commerce, how information is composed for a product, or an advertisement is essential to get positive user feedback. From infinite contents, it is important to give users information they want explicitly or implicitly. By ranking information, we can selectively provide information that satisfies users' tastes where the ranking method is actively covered in *learning to rank* [2–4, 17, 23]. Based on various *learning to rank* methods, many service platforms struggle to obtain positive user feedback by providing good content in Web searches, product recommendations, or advertisements.

In particular, *creative ranking* is recently known as a ranking approach for product advertisement. It is for selecting a creative to advertise expected to attract users' attention when composing content for a product to sell. ByteDance, a technology company operating content platforms, has improved its advertising system based on creative ranking [34]. Their proposed pre-evaluation of ad creative (PEAC) model is a pairwise ranking model using the various information of each candidate creative. By ranking creatives at the offline phase, it produces only potential creatives to their online recommender system. It has eventually improved the performance of their overall system. In addition, Alibaba, one of the largest online shopping malls, proposed another method using creative ranking [30]. They pointed out that PEAC is not flexible as it works only offline and then proposed a solution model visual-aware ranking model and hybrid bandit model (VAM–HBM) that works online. With VAM that captures visual information of creatives for ranking, they alleviated the cold start problem of typical methods based on a multi-armed bandit algorithm [28].

Meanwhile, there is a problem similar to creative ranking in e-commerce: *selecting the best from multiple images considering a topic*. As in Fig. 1, suppose that four products are on sale together within topic *'cropped sweatshirt for women'*. In this case, many e-commerce platforms expose one of the product images as the main image to give users visual information about the topic. In Fig. 1, the first image of cropped sweatshirt can be selected as the main. However, it is not easy to select the main in practice. It is hard to check whether each image matches the topic since there are often hundreds of products in a topic in real service. It is more cumbersome if there are off-topic products in the list (e.g., the fourth product in Fig. 1). In addition,

A Given Topic: *'Cropped Sweatshirt for Women'*   Selected Main Image

Fig. 1 Example of our problem: *selecting the best from multiple images considering a topic*. The main image should be selected considering (1) how attractive each image is to users and (2) how well each image fits a given topic

we should be careful not to choose a low-quality product image, even if it is suitable for the given topic (e.g., the second product in Fig. 1), as it can get less attention from users. The larger the number of products in a list, the more inefficiently time-consuming to perform the task with only human resources. Therefore, an automatic algorithm or a model is required for this problem.

Then, how do we select the best as the main image automatically? What should a model consider to pick the best? It may be suboptimal only to consider how each image catches users' attention (e.g., predicts user click-through-rate for each image) because there is no consideration of the relationship between given images and a topic. Even if the image of the pants for women in Fig. 1 is attractive enough, it should not be selected as the main because it is totally out of the given topic. A model should understand the semantic relationship between given images and a topic to make a reasonable choice. As a result, to solve the problem, we have to deal with two factors:

(F1)   *how attractive each image is to users,*
(F2)   *how well each image fits a given topic.*

The problem may be solved by existing ranking methods that predict each image's ranking score based on representations of the images and topic. We can select the best by comparing the scores. However, we experimentally observed that they do not take the factor (F2) into account sufficiently. Despite using the representation of the topic and images, they cannot guarantee that the image of women pants will never be selected as the main in Fig. 1. It is because they cannot penalize off-topic images appropriately. An additional method should be applied to overcome the limitation.

The factor (F2), consideration of the relationship between given images and topic, is closely related to the retrieval task. It is a task to search for data compatible with a query, and the search is often performed by measuring distances between embeddings. Recently, contrastive learning has shown successful performance at the retrieval task in various modalities [11, 15, 18]. Optimizing contrastive loss called InfoNCE [24] or also known as NT-Xent [5], a model minimizes distances between semantically similar embedding pairs (i.e., positive pairs) and maximizes distances between dissimilar pairs (i.e., negative pairs). Since the learning procedure is directly related to the retrieval task, contrastive learning is a key for the factor (F2). In summary, we leverage a ranking method for the factor (F1) and contrastive learning for the factor (F2). The most important thing here is to consider both factors *simultaneously*. In other words, if several images and a topic are given, a solution model should subordinate the semantic relationship between the topic and the images to its ranking scores.

In this paper, we propose a novel model CLIK (Contrastive Learning for topic-dependent Image ranKing) to solve the problem of *selecting the best from multiple images considering a topic*. It can consider both factors above by performing two significant training tasks: *topic matching* and *image ranking*, as in Fig. 2. At first, in *topic matching*, our model understands the semantic relationship between images and topics using contrastive learning inspired by CLIP [26]. Secondly, in *image ranking*, our

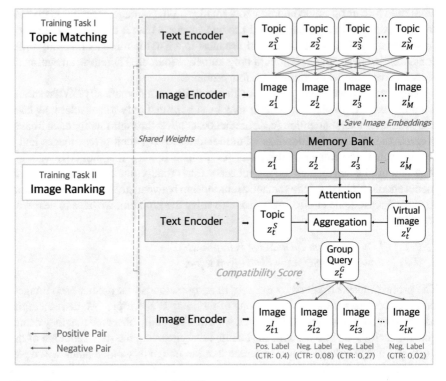

Fig. 2 Structure and training tasks of CLIK

model uses contrastive loss to rank given images considering a given topic, leveraging knowledge learned from *topic matching*. Both tasks are done simultaneously by integrated modules with shared weights. As a result, our model successfully subordinates the semantic relationship between given images and a topic to ranking scores. From the offline experiments comparing our model with several baselines, we observed that our model shows significant performance and reasonably considers the semantic relationship between given images and a topic during ranking. We also applied our model to one of the services of our platform relevant to the main problem. From online A/B testing, we got more positive feedback from users compared to baseline.

2 Related Work

2.1 Creative Ranking

In the advertising system, *creative ranking* is a method to select a creative expected to attract users' attention when composing content for a product to sell. In general, many creatives for a product are dynamically selected by online ranking algorithm. In this case, cold start problem can arise before each creative get enough impressions that are needed to make ranking result reliable. In order to treat this problem, many studies have been progressed. ByteDance proposed PEAC [34], a pairwise ranking model using representations of images, texts embedded in the images, and category information of creatives. By ranking creatives offline, it produced only reliable creatives to their own online recommender system rather than randomly sampled ones causing cold start problem. It eventually improved performance of their system. Alibaba proposed another creative ranking-based method [30]. They pointed out that PEAC of ByteDance is not flexible because it works only at the offline and then proposed VAM–HBM receiving online observations flexibly, which consists of creative ranking method and MAB algorithm. In particular, with its creative ranking part VAM capturing visual information of creatives, they alleviated cold start problem of typical MAB-based methods.

Our problem is related to creative ranking in terms of selecting the best from a given list by ranking. However, there is a difference: *consideration of a topic*. For given images and topic, we experimentally observed that general creative ranking methods cannot explicitly consider semantic relationship between images and topic well. In order to inject understanding of it into a solution model, we should use another approach additionally, not only use creative ranking.

2.2 Contrastive Learning

Intuitively, *contrastive learning* (CL) is a learning method based on data representation comparison. The goal of it is to make similar pairs closer and dissimilar pairs far apart. Unlike general supervised learning requiring human annotation, CL only needs to define a similarity distribution obtained easily through pretext tasks (e.g., data augmentation) [14, 19]. It is, therefore, one of the popular learning approaches for self-supervised learning (SSL). In computer vision, various CL-based SSL methods [5–7, 12, 31] showed successful performances for many downstream tasks outperforming supervised learning-based pre-training (e.g., pre-training with image classification on ImageNet [8]), and recent studies to apply it to other modalities are also active [16, 27, 33].

In particular, CLIP [26] shows the significance of cross-modal CL pre-training with large-scale text-image pairs collected from Websites. It learns multimodal embedding space of text and image. The authors demonstrated that CLIP performed competitively with state-of-the-art methods for many downstream tasks at computer vision and mentioned the potential for widely-applicable retrieval tasks on image, text, and image-text (cross-modal). Afterward, Jia et al. [15] proposed ALIGN using the same training procedure as CLIP but more data for training and proved that it outperformed performance in various retrieval tasks. Following the successful results, research on applying the broad utility of CLIP to the e-commerce field also proceeds [13].

We determined that CL would be appropriate for the factor (F2), that is, to understand the semantic relationship between images and topics, because CL has proven successful performance on retrieval tasks closely related to the factor (F2). Primarily, we adopted the training procedure of CLIP. Since the modality of the topic is often linguistic in the e-commerce field, training based on image-text comparison is an effective way to learn the semantic relationship between images and topics.

3 Method

In this section, we introduce our proposed CLIK. At first, we explain our problem, *selecting the best from multiple images considering a topic*. Second, we show an overview of CLIK. We describe the model structure and its two significant training tasks. At last, we explain how both tasks work in detail.

3.1 Problem Definition

The problem of *selecting the best from multiple images considering a topic* is solved by ranking given images. We assume a score exists that evaluate each image from the aspect of our problem and denote it as *a compatibility score c*. Additionally, we define $\{s, X\}$ as a group G where s is a given topic, and $X = \{x_i\}_{i=1}^{|X|}$ is a given list of images.

$$G = \{s, X\}$$
$$X = \{x_i\}_{i=1}^{|X|}$$

where the number of elements of X is greater than 1. Here, we define the modality of a topic s as text because most topic information are represented as text in many e-commerce services. As a result, our goal is to find a model f that predicts a compatibility score of each image from a given group, and then, the best image can eventually be selected as follows:

$$\{c_i\}_{i=1}^{|X|} = f(G)$$
$$c_* = \operatorname{argmax}_{c_i} \{c_i\}_{i=1}^{|X|}$$

where c_i is a compatibility score of corresponding x_i, and c_* is the compatibility score of the best image from given X. The argmax can surely be replaced as the argmin if a lower c indicates better compatibility.

For each image, the score is determined by two factors: (1) *how each image is attractive to users* and (2) *how each image fits a given topic*. For instance, suppose a topic of 'men pants for a trip to the beach' and some product images are given. In this case, the best image at least should not only be appealing enough to get users' attention but also describe the topic well. In other words, a model should predict bad compatibility scores for an image of women pants that are not fit for the given topic or an ugly image that is not attractive enough. For the ideal prediction, a model needs a reasonable label or ground truth information that indicates each image's compatibility. The problem then can be solved by various approaches depending on the labeling strategy (e.g., classification, regression).

It seems that the problem can be solved by typical ranking approaches such as *learning to rank* or creative ranking. For example, we can consider a ranking model that uses a given topic as a query and predicts the ranking scores of a given images by measuring the distance between each image and the query. In this case, does the query represent *'topic'* semantically? Experimentally, we found that it does not. It is hard to guarantee that the model understands the semantic relationship between the given topic and images well. For instance, we observed that the model could not explicitly penalize images irrelevant to the given topic. One of the challenges is that model should subordinate the semantic relationship between images and topic to compatibility scores during prediction.

3.2 Overview

The structure of CLIK is composed of dual encoders and auxiliary modules. Dual encoders are feature extractors for images and topics (see 'text encoder' and 'image encoder' in Fig. 2). Auxiliary modules include three parts: *aggregation*, *attention*, and *memory bank*. In a nutshell, they are used to generate a special query embedding for a compatibility score prediction, one of the essential components of CLIK.

Our model performs two novel training tasks. The first one is *topic matching* (*TM*). In *TM*, the model learns the semantic relationship between images and topics. The second one is *image ranking* (*IR*). In *IR*, the model predicts compatibility scores of given images considering a given topic. The best image then can eventually be selected by comparing the scores. Both tasks are done simultaneously by integrated modules with shared weights.

3.3 Topic Matching

In *topic matching* (*TM*), CLIK understands the semantic relationship between various images and topics. Inspired by CLIP [26], a significant cross-modal contrastive learning model, we adopted its training procedure. For given M pairs composed of image and corresponding topic, the model predicts M correct pairs from M^2 possible pairs that include $M \times (M - 1)$ incorrect pairs. Since the representations of images and topics are only needed, crowd-sourced labels are not required. Therefore, we can use a large amount of data efficiently with no limitation of supervision. To optimize the following L_{matching}, which is the same as NT-Xent loss [5], CLIK maximizes the similarity of correct pairs and minimizes that of the others.

$$L_{\text{matching}} = (L_{\text{S2I}} + L_{\text{I2S}})/2$$

$$L_{\text{S2I}} = -\frac{1}{M}\sum_{m=1}^{M} \log \frac{\exp\left(\text{sim}(z_m^S, z_m^I)/\tau\right)}{\sum_{i=1}^{M}\exp\left(\text{sim}(z_m^S, z_i^I)/\tau\right)}$$

$$L_{\text{I2S}} = -\frac{1}{M}\sum_{m=1}^{M} \log \frac{\exp\left(\text{sim}(z_m^I, z_m^S)/\tau\right)}{\sum_{i=1}^{M}\exp\left(\text{sim}(z_m^I, z_i^S)/\tau\right)}$$

where z_m^I and z_m^S denote an image and topic embedding of the mth group G_m in mini-batch, τ is a temperature parameter, and $sim(\cdot, \cdot)$ is a cosine similarity function. The dimension of all embeddings above is the same.

To compose a mini-batch, we make M pairs of an image and corresponding topic from M groups. A positive pair is made by sampling an image from each group and pairing it with its corresponding topic, and the images and topics that do not match are all regarded as negative pairs (see the 'topic matching' part in Fig. 2). Comparing various images and topics, CLIK (especially dual encoders) learns embedding space which reflects the semantic relationship between images and topics. Leveraging the space, in the other training task *image ranking*, our model then subordinates the semantic relationship between given images and topic to compatibility scores.

3.4   Image Ranking

In *image ranking* (*IR*), for a given group, CLIK predicts the compatibility score of each image considering the representation of the given images and a topic. Then, we can select the best image by comparing the scores. Our model performs metric learning using contrastive loss over cosine similarity between given images and a special query embedding. Optimizing the following loss function L_{ranking}, the model makes compatible images closer to the query and the others farther away from it.

$$L_{\text{ranking}} = -\frac{1}{N}\sum_{i=1}^{N} \log \frac{\exp\left(\text{sim}(z_i^G, z_{i*}^I)/\tau\right)}{\sum_{k=1}^{K}\exp\left(\text{sim}(z_i^G, z_{ik}^I)/\tau\right)}$$

where z_i^G is a special query embedding of the ith group G_i of a mini-batch, z_{i*}^I is the best image embedding out of K included image embeddings $\{z_{ik}^I\}_{k=1}^{K}$ from X_i, and N is a mini-batch size. A mini-batch is composed by sampling K images from N groups. The dimension of all embeddings above is the same.

For each sampled group, we label the most relatively compatible image among K images as positive and the others as negative. Comparing cosine similarity between images and the query z^G, CLIK classifies the most compatible image from given K images. We then regard the cosine similarity $sim(z^G, z_k^I)$ as a compatibility score c_k of x_k.

3.4.1 Group Query

Group query z^G is a special query embedding representing the overall information of a group G. It is one of the essential elements of CLIK as it helps our model successfully perform two training tasks simultaneously.

Until we adopted *group query*, we considered using a given topic z^S as a query. In this case, however, the performance was closely the same as random ranking that just randomly shuffles the given image list. We guessed that the cause of this disaster lies in *the pairing contradiction problem* between two training tasks. Since both tasks depend on the same distance metric (cosine similarity) between embedding pairs, there is a risk of collision when the model performs both tasks simultaneously. The pair in *TM* consists of various images and corresponding topics (i.e., 'image↔topic'). In *IR*, for a given group, if we adopt an embedding of a given topic as a query, compatibility scores will be defined as the distance between image and topic, the same composition as *TM*.

Due to the sameness, a collision occurs. Since *TM* has an inter-group characteristic, pairs between images, and a topic from the same group are tentatively labeled as positive. On the contrary, since *IR* has an intra-group characteristic, only one pair that includes the most compatible image is labeled as positive for a given group. This discrepancy prevents CLIK from learning appropriate solution space. For instance, there are many cases where an image pulls to its corresponding topic in *TM* (labeled as positive) but pushes away from it in *IR* (labeled as negative). For this reason, the key to using *Group Query* embedding is to overcome the pairing contradiction. We found that CLIK performs both training tasks successfully with *group query*, which means that CLIK eventually subordinates the semantic relationship between given images and a topic to compatibility scores, one of the challenges for our problem.

A *group query* embedding z^G is generated based on the auxiliary modules, aggregating a given topic z^S and another special embedding called *virtual image* embedding z^V (see the 'image ranking' part in Fig. 2).

$$z^G = Aggregation(z^S, z^V)$$

where *virtual image* embedding z^V is an embedding of a virtual image that semantically fits a given topic. An attention mechanism generates it. For a group G, the attention operation is performed by using the given topic z^S as a query and '*memory bank*', one of the auxiliary modules, as both keys and values as follows:

$$z^V = \sum_j \alpha_j z_j^I, \quad z_j^I \in Memory\ Bank$$

$$\{\alpha_j\}_{j=1}^M = \text{Softmax}\left(z^S \odot Memory\ Bank\right)$$

$$Memory\ Bank = \left\{z_j^I\right\}_{j=1}^M$$

where *memory bank* stores memories of various images that CLIK has encountered. In *TM*, the model meets numerous images sampled from many groups. The image embeddings from *TM* are stored explicitly in the *memory bank* and then are used to generate *group query* embedding. As model parameters are updated, we update the bank with newly extracted image embeddings for every training step to prevent the problem of stored embeddings being outdated [12]. We reported the ablation study for using *Group Query* embedding in detail in the experiment section.

3.5 Summary

To solve the main problem, CLIK performs two training tasks simultaneously: *topic matching* and *image ranking*. In *topic matching*, the model understands the semantic relationship between various images and topics. Optimizing L_{matching}, the model learns to determine which image matches which topic in semantic aspect. In *image ranking*, optimizing L_{ranking}, the model selects the most relatively compatible image from a given group by predicting a compatibility score for each image. Unlike typical ranking methods, CLIK predicts the scores considering the representation of given images and the semantic relationship between the images and the topic by leveraging knowledge learned from *topic matching*. As a result, CLIK minimizes loss function L_{CLIK} as follows:

$$L_{\text{CLIK}} = L_{\text{matching}} + \lambda \cdot L_{\text{ranking}}$$

where λ is a scalar to adjust the contribution of two loss functions. We set it to 20.

4 Experiments

In this section, we conduct experiments on a real-world dataset to evaluate CLIK. We reported offline and online result based on *online special exhibition*, one of our services. At first, we explain *online special exhibition* and how we collected data from the service. Secondly, we show offline evaluation results. Lastly, we show online evaluation results.

4.1 Online Special Exhibition

Online special exhibition is a service that collects and sells products suitable for a special theme. On the main page, users can grasp at a glance the theme as in Fig. 3. Each theme is described through not only text information such as title and category but also a representative image. Especially, the representative image has

Fig. 3 Example of *online special exhibition*. The first part is overall information of an exhibition, and the others are products of the exhibition. For CLIK, text embedded in the exhibition is used as *a topic*, product images are used as *a list of images*, and the exhibition is regarded as *a group*

been determined by our service operators recently. Since there often exist products that are off-themed from the corresponding exhibition or have low image quality, the operators should filter them delicately. This human-based process has problems in that personal tastes are subordinated to the selection, and it is inefficient to pick the representatives for hundreds of exhibitions daily. To overcome the problems, we tried to apply CLIK to the so-called '*representative image selection*'. With CLIK, we can make automatic selections based on the estimated potential of each image as a representative image.

Figure 4 is an explanation for the application of CLIK. We regard an exhibition as *a group G*, the product images as *a list of images X*, and the text describing its theme as *a topic s*. In addition, since we collect various implicit user feedbacks on each product, we adopt one of them as an indicator or criterion for labeling to guide CLIK infer *a compatibility score c* for each image. For example, each product's user click-through rate (CTR) can be adopted as the criterion. In this case, for a given exhibition, the image of a product whose CTR is the greatest among the given products is labeled as positive, and the others are labeled as negative in *IR*. Note that user feedback generated from a service relevant to our problem can only be adopted as the criterion. For example, the user CTR of products from service with no topical information covering some products cannot be adopted as a criterion because the service is far from our problem.

In offline evaluation, we evaluated how accurately CLIK predicts the representative image based on two datasets. We adopted two metrics suitable for *representative image selection*. Additionally, we observed how user feedback changes by applying CLIK online. As a result, we obtained successful results in both evaluations.

4.2 Data Collection

We collected *online special exhibition* dataset from August to November 2021. For labeling, we conducted the collection process based on two labeling criteria: CTR and review count (i.e., the number of user reviews) for each product generated from *online special exhibition*. Consequently, we collected two types of datasets.

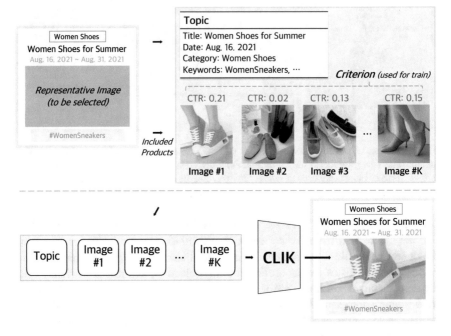

Fig. 4 *Representation image selection* by CLIK in *online special exhibition*

Table 1 Statistics of data collected from *online special exhibition* service

Type	Criterion	# Exhibitions	# Products	Date
1	CTR	1605	104,716	Aug. 2021–Nov. 2021
2	Review count	4174	293,501	Aug. 2021–Nov. 2021

Since some products are banned from sale or deleted, we dropped the exhibitions where less than 50% of the registered products are collected. For the CTR criterion dataset (Type 1 at Table 1), we removed exhibitions with no clicks and just collected products whose impressions are 10 or more to use only products with reliable CTR. Additionally, we dropped the exhibitions where the uniqueness of CTR is less than 2, and the number of zero-CTR products is more than 5. On the one hand, for the dataset using the review count criterion (Type 2 at Table 1), we dropped exhibitions with no reviews and only collected products whose reviews were 10 or more. Also, we excluded the exhibitions where the uniqueness of review counts is less than 2, and the number of zero-review products is more than 5. Since the image sizes of products varied (e.g., 600 × 600, 1000 × 1000), we pre-resized all images for training efficiency. We collected title, keywords, publication date, and category information for topic information. Table 1 describes statistics of collected data. After the collection, we used 90% of them as train dataset and the other 10% as test dataset by random sampling.

4.3 Experiment Settings

4.3.1 Metrics

Since the main goal of our problem is to select the best image, we adopted mean reciprocal rank (*MRR*) and newly defined TopK-Top1 accuracy (*TopK-Top1*) as evaluation metrics to focus on the first-ranked image.

$$MRR = \frac{1}{N} \sum_{n=1}^{N} \frac{1}{Rank(x_{n*}, f(G_n))}$$

$$TopK\text{-}Top1 = \frac{1}{N} \sum_{n=1}^{N} \mathbf{1}(f(G_n), K)$$

where $Rank(x_i, f(G))$ is a predicted rank of image x_i by a model f for a given group G, and $\mathbf{1}(f(G, K))$ is an indicator function that outputs 1 if the true rank of the image predicted as the best is less than K and 0 for the other cases. In short, *MRR* is for observing how the model predicts the actual first-ranked image, and *TopK-Top1* is for observing the actual rank of the image predicted as the best. The best and worst value of both metrics is 1 and 0.

4.3.2 Implementation Details

We adopt BERT [9] to encode topics and ViT [10] to encode images. They are known as scalable and efficient models using transformer architecture [29]. We used BERT composed of 6 layers with 768 hidden dimension, ViT composed of 12 layers with 384 hidden dimension, and initialized them with pre-trained weights. The dimension of output embeddings from both encoders was set to 128. We adopted a dot product-based attention module for *attention*, a one-layered fully connected layer as *aggregation* that maps concatenated embeddings of a topic embedding and *virtual image* embedding into a *group query* embedding of dimension 128. We added a tensor buffer for *memory bank* into our model by referencing MoCo [12].

We apply only a random crop to the input image with a size of 224×224 for the train and only resize it to the same size for the evaluation. We pre-processed text information of topics in '[CLS] *Title* [SEP] *Publication Date* [SEP] *Category* [SEP] *Keywords* [SEP]', the compatible input format for BERT. The maximum length of the text is defined as 128 with filling empty spaces with '[PAD]' tokens.

In *TM*, we set the batch size to 512 ($M = 512$), and the size of *memory bank* is the same. In *IR*, we set the number of sampled groups to 12 ($N = 12$), and each group consists of randomly sampled 20 images ($K = 20$) for the train. On the one hand, we set N to 1 and K to 50 for evaluation in *IR*, the numbers that better reflect our real online service. We use AdamW [21] applying weighted decay regularization to all weights with a decay rate of 0.1. We update the topic encoder (BERT) with an initial learning rate of 0.00005 and 0.0001 for the other parts and incorporate learning rate warm-up over the first 1% steps, followed by cosine decay [20] to lower the learning rate to zero. The temperature parameter τ is fixed as 0.07. We build our model and experiment settings based on PyTorch [25], a popular deep learning framework, and use the automatic mixed-precision [22] to accelerate training and save memory. We trained our model for 10 epochs with 4 P40 GPUs.

4.3.3   Baselines

To verify the significance of CLIK, we compare it with a few loss functions: *triplet*, *pairwise*, and *pointwise* loss. They are general loss functions for *learning to rank* or creative ranking. We compared performance between the baselines and our model. Additionally, we observed the inference results to evaluate whether each model considers the semantic relationship between images and a given topic well for ranking.

- **Triplet Loss** With triplet loss, a model takes an anchor, a positive, and a negative. Then, the model makes the positive closer to the anchor and the negative farther away than a margin. For a given group G, we use a topic s as an anchor and assign positive and negative to two randomly sampled images from X by comparing their values of the pre-defined criterion (e.g., CTR).

$$L_{\text{triplet}} = \max\left(\|z^S - z^I_{\text{pos}}\|^2 - \|z^S - z^I_{\text{neg}}\|^2 + \alpha, 0 \right)$$

  where z^S is an embedding of a given topic, z^I_{pos} and z^I_{neg} are positive and negative product image sampled from group G, and α is the margin set to 0.2. We then regard the distance between embeddings of a topic s and an image x as a compatibility score c of x.
- **Pairwise Loss** Pairwise loss is optimized by comparing a pair of samples as in [2, 34]. For our problem, we randomly sample two images from a group G at first. Then, extract embeddings of each image and given topic s by corresponding encoders and concatenate each image embedding with the topic embedding. By forwarding both concatenated embeddings to a one-layered fully connected layer, model predicts score for each sampled image. The score then used for comparison for pairwise loss, and we regard it as a compatibility score c.
- **Pairwise Loss** A model with pairwise loss optimizes the loss by comparing a pair of samples as in [2, 34]. For our problem, we randomly sample two images from group G. Then, we extract embeddings of each image and given topic s by corresponding encoders and concatenate each image embedding with the topic embedding. By forwarding both concatenated embeddings to a one-layered fully connected layer, the model predicts a score for each sampled image. The score is then used for comparison during optimization, and we regard it as a compatibility score c.

$$L_{\text{pairwise}} = - \left(y \log \sigma (c_i - c_j) + (1 - y) \log(1 - \sigma(c_i - c_j)) \right)$$

where c_i is a compatibility score of x_i, σ is a sigmoid function, and y is 1 if the value of the criterion of image x_i is greater than that of image x_j and 0 for the other case.

- **Pointwise Loss** To optimize pointwise loss, a model predicts scores of samples one by one. It is generally optimized by minimizing mean squared error between labels and predicted scores. Due to the hardness of actual value prediction, this approach is known to have lower performance than the pairwise loss that considers the only relative relationship of a pair [34]. We extract embeddings of each image and its corresponding topic and concatenate them. Then, predict a score by forwarding the embedding through a one-layered fully connected layer. We define the criterion value for each product image as a label and regard the predicted score as compatibility score c.

$$L_{\text{pointwise}} = \frac{1}{N} \sum_{n=1}^{N} \left(y_{nj} - c_{nj} \right)^2$$

where y_{nj} and c_{nj} are a value of criterion and a compatibility score for x_{nj} from group G_n.

For the baselines above, we set the same dual encoders and the same dimension of embeddings as CLIK for fairness (i.e., BERT and ViT for dual encoders and 128 for encoded embedding dimension).

4.4 Offline Evaluation

4.4.1 Comparison with Baselines

We compared CLIK with baselines using two types of datasets where one uses CTR, and the other uses review count as a labeling criterion. According to the Table 2,

Table 2 Offline evaluation compared to baselines

	CTR				Review count			
	MRR	Top1-Top1	Top3-Top1	Top5-Top1	MRR	Top1-Top1	Top3-Top1	Top5-Top1
CLIK	**0.1226**	**0.0496**	0.0729	0.1283	0.1627	**0.0828**	**0.1379**	**0.2103**
Triplet	0.102	0.0233	0.0758	0.1254	**0.1645**	0.0448	0.1000	0.1414
Pairwise	0.1063	0.0379	0.0641	0.1195	0.1380	0.0448	0.0828	0.1310
Pointwise	0.121	0.0379	**0.0947**	**0.1457**	0.1078	0.0207	0.0517	0.0724
Random	0.0899	0.02	0.06	0.1	0.0899	0.02	0.06	0.1

CLIK shows significant performance overall compared to the baselines. We could conclude that CLIK is an especially suitable method for the *representative image selection*. The first-ranked one is more important than the others because *Top1-Top1* accuracy is superior to the others. In addition, since the overall *TopK-Top1* accuracy is relatively high, CLIK is likely to predict at least a high-ranked image as the first more stably, even if it is not a first-ranked image.

In addition, we can see the vital characteristic of CLIK from the inference result comparison. Figure 5 shows ranking results for a given product list of an exhibition, including 50 products. From the text, we can guess that the given exhibition's theme is relevant to pants for men (e.g., 'Title: Men Trousers, Bending, Spandex Pants'). According to the topic, the representative should visually depict pants for men. Therefore, the model should not rank images of given top products high, or the compatibility scores of images for the top products should be lower than those of bottom products. From this aspect, CLIK does its job much better than the baselines. According to the best-ranked 10 images in Fig. 5, there are no top product images from CLIK, while the baseline includes several top products. Additionally, since the worst-ranked 10 images from CLIK are mainly composed of top product images, we can conclude that our model effectively subordinates the semantic relationship between given images and topic to the compatibility scores. On the other hand, since the inference result of the baseline shows a randomly mixed top and pants in worst-ranked and best-ranked 10 images, it seems that general ranking methods cannot capture the semantic relationship.

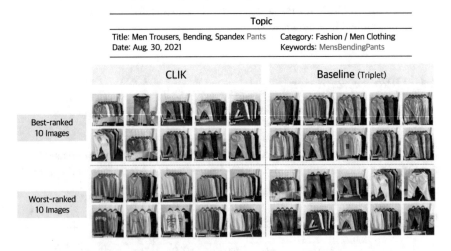

Fig. 5 Inference result comparison between CLIK and the baseline with triplet loss. Other baselines show similar inference patterns to those of triplet loss

Table 3 Experiment for query usage (criterion: review count)

Query	MRR	Top1-Top1	Top3-Top1	Top5-Top1
Group query	**0.1627**	**0.0828**	**0.1379**	**0.2103**
Virtual image	0.1232	0.0483	0.1103	0.1655
Topic	0.0909	0.0207	0.0621	0.1000
Random	0.0899	0.02	0.06	0.1

4.4.2   Usage of Group Query

Group query embedding is one of the essential elements of CLIK. It helps the model avoid *the pairing contradiction problem* between two training tasks. For a given group, we could consider using a given topic as a query ('topic' in Table 3) until we adopted *group query*. In this case, however, the performance was similar to random ranking, which randomly shuffles the given image list ('random' in Table 3). On the other hand, according to the superior result of *group query* ('group query' in Table 3) compared to the case of *Topic*, we conclude that it is a key to overcoming the contradiction problem. With the new modality of *group query* combining images and topic, each pair composition of both tasks becomes different, and the model can eventually perform both tasks successfully simultaneously. Meanwhile, we tested an additional hypothesis. We expected that the *virtual image* embedding for generating *group query* could also prevent the contradiction problem in the same way as in the case of *group query*. From the result of *virtual image* in Table 3, we found that the performance is superior to the case of *topic* even in this case. However, it is worse than the case of *group query* where we can conclude that the combined information inherent in *group query* is helpful for CLIK.

4.5   *Online Evaluation*

We analyzed the effect of CLIK in the real world through online A/B testing at our service *online special exhibition*. Currently, in our service, *representative image selection* is just made by randomly selecting one of the product images of each exhibition as the representative to use no human resources. Therefore, for the test, we compared CLIK with all our service users' random selection process tracking conversion rate (CVR). We conducted the test for 11 days.

The result is in Fig. 6, where 'baseline' is the original random selection process. We measured users' CVR with min-max normalization. On all days except one day during the test, normalized CVR was higher in applying CLIK than in the other case. In particular, we saw an overall improvement of about 44%, demonstrating CLIK's ability to produce good content for users with the sophisticated consideration required for *online special exhibition*. We are actively applying CLIK to our service based on the successful results.

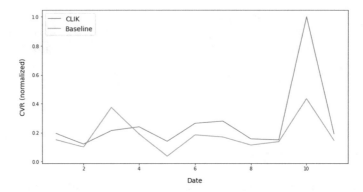

Fig. 6 Online A/B testing result. In the case of baseline, representative image for each exhibition is randomly selected from its product list. It shows an overall improvement of about 44% for user CVR when we apply CLIK to our live service than the case of baseline

5 Conclusion

In this paper, we proposed CLIK for the problem of *selecting the best from multiple images considering a topic*. With two training tasks, our model solves the problem by understanding how each image is attractive to users, and how each image fits a given topic. We demonstrated that CLIK is superior to existing ranking methods and encourages positive feedback in our live service.

Despite the significance, our work has a limitation. We use one of the values of each image as a labeling indicator for ranking. Still, it is hard to guarantee that the value is determined only by the image's appearance or the relationship between the image and a given topic. For instance, when we adopt the CTR of each product as its value, CTR may be high not because the product image is attractive but only because of a special event at the service. Thus, reasonable refinement is required to train the model ideally. That is, we must rule out factors outside the assumptions of our problem to find a reliable value. Meanwhile, we further expect CLIK to be compatible with various modalities. Inspired by CLIP [26], CLIK only deals with texts and images. Since many contrastive learning approaches using various modalities have been proposed recently [1, 27, 32], we believe we can improve CLIK to perceive various modalities in future. The modality-agnostic model will be of greater help in various e-commerce services.

In the e-commerce field, although we frequently face situations similar to *selecting the best from multiple images considering a topic*, solutions for them have not been actively studied. It is not easy to solve them optimally just with a general ranking approach because it does not consider many factors of the situations. We hope this work motivates future research to tackle these problems in various research groups or e-commerce platforms.

References

1. Akbari H, Yuan L, Qian R, Chuang WH, Chang SF, Cui Y, Gong B (2021) Vatt: transformers for multimodal self-supervised learning from raw video, audio and text. Adv Neural Inf Process Syst 34:24206–24221
2. Burges C, Shaked T, Renshaw E, Lazier A, Deeds M, Hamilton N, Hullender G (2005) Learning to rank using gradient descent. In: Proceedings of the 22nd international conference on Machine learning, pp 89–96
3. Burges CJ (2010) From ranknet to lambdarank to lambdamart: an overview. Learning 11(23–581):81
4. Cao Z, Qin T, Liu TY, Tsai MF, Li H (2007) Learning to rank: from pairwise approach to listwise approach. In: Proceedings of the 24th international conference on Machine learning, pp 129–136
5. Chen T, Kornblith S, Norouzi M, Hinton G (2020a) A simple framework for contrastive learning of visual representations. In: International conference on machine learning, PMLR, pp 1597–1607
6. Chen X, Fan H, Girshick R, He K (2020b) Improved baselines with momentum contrastive learning. arXiv:2003.04297
7. Chen X, Xie S, He K (2021) An empirical study of training self-supervised vision transformers. In: Proceedings of the IEEE/CVF international conference on computer vision, pp 9640–9649
8. Deng J, Dong W, Socher R, Li LJ, Li K, Fei-Fei L (2009) Imagenet: a large-scale hierarchical image database. In: 2009 IEEE conference on computer vision and pattern recognition, IEEE, pp 248–255
9. Devlin J, Chang MW, Lee K, Toutanova K (2018) Bert: pre-training of deep bidirectional transformers for language understanding. arXiv:1810.04805
10. Dosovitskiy A, Beyer L, Kolesnikov A, Weissenborn D, Zhai X, Unterthiner T, Dehghani M, Minderer M, Heigold G, Gelly S, et al. (2020) An image is worth 16x16 words: transformers for image recognition at scale. arXiv:2010.11929
11. Gao T, Yao X, Chen D (2021) Simsce: simple contrastive learning of sentence embeddings. arXiv:2104.08821
12. He K, Fan H, Wu Y, Xie S, Girshick R (2020) Momentum contrast for unsupervised visual representation learning. In: Proceedings of the IEEE/CVF conference on computer vision and pattern recognition, pp 9729–9738
13. Hendriksen M, Bleeker M, Vakulenko S, Noord Nv, Kuiper E, Rijke Md (2022) Extending clip for category-to-image retrieval in e-commerce. In: European conference on information retrieval. Springer, pp 289–303
14. Jaiswal A, Babu AR, Zadeh MZ, Banerjee D, Makedon F (2020) A survey on contrastive self-supervised learning. Technologies 9(1):2
15. Jia C, Yang Y, Xia Y, Chen YT, Parekh Z, Pham H, Le Q, Sung YH, Li Z, Duerig T (2021) Scaling up visual and vision-language representation learning with noisy text supervision. In: International conference on machine learning, PMLR, pp 4904–4916
16. Jiang D, Li W, Cao M, Zou W, Li X (2020) Speech simclr: combining contrastive and reconstruction objective for self-supervised speech representation learning. arXiv:2010.13991
17. Karmaker Santu SK, Sondhi P, Zhai C (2017) On application of learning to rank for e-commerce search. In: Proceedings of the 40th international ACM SIGIR conference on research and development in information retrieval, pp 475–484
18. Krishna T, McGuinness K, O'Connor N (2021) Evaluating contrastive models for instance-based image retrieval. In: Proceedings of the 2021 international conference on multimedia retrieval, pp 471–475
19. Le-Khac PH, Healy G, Smeaton AF (2020) Contrastive representation learning: a framework and review. IEEE Access 8:193907–193934
20. Loshchilov I, Hutter F (2016) SGDR: stochastic gradient descent with warm restarts. arXiv:1608.03983

21. Loshchilov I, Hutter F (2017) Decoupled weight decay regularization. arXiv:1711.05101
22. Micikevicius P, Narang S, Alben J, Diamos G, Elsen E, Garcia D, Ginsburg B, Houston M, Kuchaiev O, Venkatesh G, et al (2017) Mixed precision training. arXiv:1710.03740
23. Mishra S, Verma M, Zhou Y, Thadani K, Wang W (2020) Learning to create better ads: Generation and ranking approaches for ad creative refinement. In: Proceedings of the 29th ACM international conference on information & knowledge management, pp 2653–2660
24. Van den Oord A, Li Y, Vinyals O (2018) Representation learning with contrastive predictive coding. arXiv:1807.03748
25. Paszke A, Gross S, Massa F, Lerer A, Bradbury J, Chanan G, Killeen T, Lin Z, Gimelshein N, Antiga L, et al (2019) Pytorch: an imperative style, high-performance deep learning library. In: Advances in neural information processing systems, vol 32
26. Radford A, Kim JW, Hallacy C, Ramesh A, Goh G, Agarwal S, Sastry G, Askell A, Mishkin P, Clark J, et al (2021) Learning transferable visual models from natural language supervision. In: International conference on machine learning, PMLR, pp 8748–8763
27. Shin K, Kwak H, Kim SY, Ramstrom MN, Jeong J, Ha JW, Kim KM (2021) Scaling law for recommendation models: towards general-purpose user representations. arXiv:2111.11294
28. Slivkins A (2019) Introduction to multi-armed bandits. arXiv:1904.07272
29. Vaswani A, Shazeer N, Parmar N, Uszkoreit J, Jones L, Gomez AN, Kaiser L, Polosukhin I (2017) Attention is all you need. CoRR abs/1706.03762, http://arxiv.org/abs/1706.03762
30. Wang S, Liu Q, Ge T, Lian D, Zhang Z (2021) A hybrid bandit model with visual priors for creative ranking in display advertising. In: Proceedings of the web conference 2021, pp 2324–2334
31. Wu Z, Xiong Y, Yu SX, Lin D (2018) Unsupervised feature learning via non-parametric instance discrimination. In: Proceedings of the IEEE conference on computer vision and pattern recognition, pp 3733–3742
32. Ye R, Wang M, Li L (2022) Cross-modal contrastive learning for speech translation. arXiv:2205.02444
33. Yue Z, Wang Y, Duan J, Yang T, Huang C, Tong Y, Xu B (2021) Ts2vec: towards universal representation of time series. arXiv:2106.10466
34. Zhao Z, Li L, Zhang B, Wang M, Jiang Y, Xu L, Wang F, Ma W (2019) What you look matters? offline evaluation of advertising creatives for cold-start problem. In: Proceedings of the 28th ACM international conference on information and knowledge management, pp 2605–2613

A Dataset for Learning Graph Representations to Predict Customer Returns in Fashion Retail

Jamie McGowan, Elizabeth Guest, Ziyang Yan, Cong Zheng, Neha Patel, Mason Cusack, Charlie Donaldson, Sofie de Cnudde, Gabriel Facini, and Fabon Dzogang

Abstract We present a novel dataset collected by ASOS (a major online fashion retailer) to address the challenge of predicting customer returns in a fashion retail ecosystem. With the release of this substantial dataset, we hope to motivate further collaboration between research communities and the fashion industry. We first explore the structure of this dataset with a focus on the application of Graph Representation Learning in order to exploit the natural data structure and provide statistical insights into particular features within the data. In addition to this, we show examples of a return prediction classification task with a selection of baseline models (i.e. with no intermediate representation learning step) and a graph representation based model. We show that in a downstream return prediction classification task, an F1-score of 0.792 can be found using a Graph Neural Network (GNN), improving upon other models discussed in this work. Alongside this increased F1-score, we also present a lower cross-entropy loss by recasting the data into a graph structure, indicating more robust predictions from a GNN-based solution. These results provide evidence that GNNs could provide more impactful and usable classifications than other baseline models on the presented dataset, and with this motivation, we hope to encourage further research into graph-based approaches using the ASOS GraphReturns dataset.

Keywords Graph representation learning · Neural message passing · Edge classification · Fashion retail dataset · Customer return prediction

1 Introduction

Part of the unique digital experience that many fashion retailers deliver is the option to return products at a small or no cost to the customer. However, unnecessary ship-

J. McGowan · E. Guest · Z. Yan · C. Zheng · G. Facini
Department of Physics and Astronomy, University College London, London, UK

N. Patel · M. Cusack · C. Donaldson · S. de Cnudde · F. Dzogang (✉)
ASOS.com, London, UK
e-mail: fabon.dzogang@asos.com

© The Author(s), under exclusive license to Springer Nature Switzerland AG 2023
H. J. Corona Pampín and R. Shirvany (eds.), *Recommender Systems in Fashion and Retail*,
Lecture Notes in Electrical Engineering 981,
https://doi.org/10.1007/978-3-031-22192-7_6

ping of products back and forth incurs a financial and environmental cost. With many fashion retailers having a commitment to minimising the impact of the fashion industry on the planet, providing a service which can forecast returns and advise a customer of this at purchase time is in line with these goals.

With the continual development of e-commerce platforms, it is important that systems are able to model the user's preferences within the platform's ecosystem by using the available data to guide users and shape the modern customer experience. One approach to this challenge, which has sparked huge interest in the field of recommendation systems [15], is representation learning-based methods. Representation learning provides a framework for learning and encoding complex patterns present in data, which more naive machine learning (ML) approaches are unable to capture as easily. However at present, the available data that is able to facilitate such research avenues is scarce. Further to this, the number of available datasets which include anonymised customer and product information (and their interactions) is even less available.

E-commerce platforms in the fashion industry are in a unique position to contribute to this research by making data publicly available for use by the machine learning community. Of particular interest to ASOS is the application of machine learning to predicting customer returns at purchase time; due to this, we present the ASOS GraphReturns dataset in this article. The labelled purchase (return or not returned) connections between customers and products in this dataset naturally lends itself to a graph structure which has motivated our interest in encouraging the exploration of graph representation learning-based solutions, which we provide an example of in Sect. 4. Graph neural networks (GNNs) have been the subject of immense success in recent years [3, 4, 10, 13, 14] and provide an intuitive way to exploit structured data. Another benefit of using GNNs is that they are able to make predictions for new instances not seen before. This is a particular attractive feature for industry environments where new products and customers are continually added.

In this work, we first present the ASOS GraphReturns dataset[1] and discuss some of the properties and features of this data. Using this data, we then provide some examples demonstrating the use of GNNs with this data based on the downstream task of predicting customer returns. This information may then be used to inform customers based on their choice and make a personalised recommendation (i.e. a different size, style, colour etc.) at purchase time that has a lower probability of being returned.

The structure of the document is as follows: Sect. 2 describes the novel fashion retail dataset, Sect. 3 overviews the methodology, and some example benchmark results are discussed in Sect. 4. Finally in Sect. 5, we summarise this contribution and provide some insights into potential further studies which could benefit from this dataset.

---

[1] The dataset can be found at https://osf.io/c793h/.

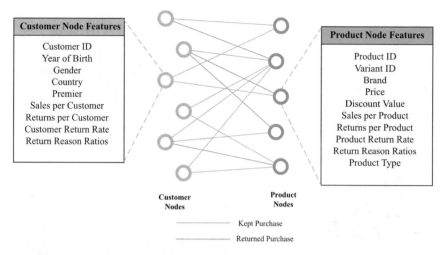

Fig. 1 The raw data structure includes customer and product specific information linked by purchases. These purchase links are labelled with a no return (blue) or return (red) label. The entire list of node features for customers and products is also provided here

2 Data Description

The train (test) data contains purchases and returns recorded by ASOS between Sept–Oct 2021 (Oct–Nov 2021), including the corresponding anonymous customer and product variant[2] specific information. The data is organised into customers (with hashed customer IDs to preserve anonymity), product variants and events (i.e. a purchase or return of a product by a customer). The training (testing) dataset includes ~770,000 (~820,000) unique customers and ~410,000 (~410,000) product variants, where every customer has at least one return and each product variant has been purchased at least once. To connect customers and products the data contains a total of 1.4 M (1.5 M) purchase events each labelled as a return (1) or no return (0) in both the training and testing datasets. The problem of predicting customer returns is then presented as an edge classification task as depicted in Fig. 1. This structure is similar to that of the Amazon reviews data [9] which also includes labelled links between customers and products.

Within each customer/product variant node, we also include specific node features, such as the average return rate, the ratios of different reasons for returns, and historical information relating to the number of purchases/returns made. Figure 1 displays an exhaustive list of all the features included in this dataset. Figure 2 (left) displays a subset of correlations between customer (top) and product (bottom) features. Within these correlations, one can observe strong associations such as male customers being less likely to make a return or a more expensive product in general

---

[2] Note that product variants include variations in size and colour and therefore a product may contain multiple variants.

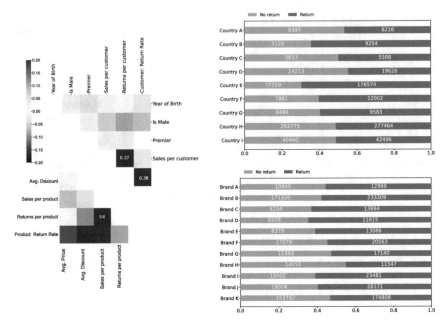

Fig. 2 General summary of data statistics including correlations between customer and product specific features (left) and distributions of return labels (right) within each country (top) and brand (bottom)

having a higher return rate. Figure 2 (right) summarises a selection of statistics related to the distribution of return labels across countries and brands included within the data. It can be seen that the data shows a larger proportion of returns across specific individual markets which could prove useful in ML-based classification tasks.[3]

Of particular interest to neural message passing techniques is the inherent graph structure that this dataset holds. In order to apply graph neural networks to data, one must first arrange the data into nodes that contain specific features and edges that link these node instances. This extra potential structure that can be constructed from the ASOS GraphReturns dataset further enhances the modality of the data from the raw structure and node features/correlations discussed above. In Fig. 3, we show the data in an undirected heterogeneous graph structure with five different edge types linking customers to their shipping countries and product variants to each other and their corresponding brands, product types and top return reasons by defining intermediate virtual nodes in all cases. These virtual nodes can be constructed in multiple ways; however, in this paper, the virtual nodes contain an averaged set of features for each instance; i.e. a product-type node will contain the average set of feature values for all products linked to this node.

---

[3] Due to the manner in which this dataset is constructed (i.e. only including customers who have at least one return), these statistics do not reflect the true ASOS purchase/return statistics.

3 Methodology

In this section, we present the methodology for a number of example baseline methods applied to the task of predicting customer returns in Sect. 4. The methods considered here aim to provide an early benchmark for future studies involving this dataset. For the graph representation learning-based approach, the data is arranged into a highly connected structure with virtual nodes for: customer shipping countries, products, product types, product brands and top return reasons for product variants as described in Fig. 3.

We investigate the use of a logistic regression, a 2-layer MLP, a random forest [1] and an XGBoost [2] classifier trained directly on the raw data (i.e. not arranged into a graph) described in Sect. 2. For these models, the customer and product specific features are joined by each labelled purchase link in the data. Further to this, we also investigate a benchmark for a GNN-based model trained in conjunction with the same baseline 2-layer MLP as a classifier head. In this case, the output of the GNN is the learnt embeddings and the MLP provides a final classification layer for the downstream tasks.

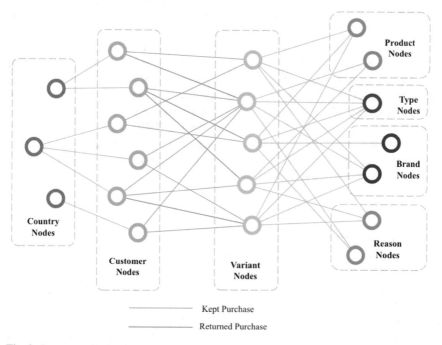

Fig. 3 Representation of the richer graph structure contained within the ASOS returns data and how it can be recast into a form better suited to graph representation learning. Virtual nodes are shown for countries, products, product types, brands and return reasons with extra connections added to each customer and product variant node

To construct an embedding for an edge $\mathbf{e}_{ab}$ between two nodes a and b, in general, one can perform an operation involving both representations for each node,

$$\mathbf{e}_{ab} = \mathcal{O}\left(\mathbf{h}_a^{(K)}, \mathbf{h}_b^{(K)}\right). \tag{1}$$

where in the case described above, $\mathcal{O}$ is described as a two-layer MLP classifier which performs the final classification from the output of the GNN.

The output of the MLP classifier head is then the predicted probability for the two class labels (return or no return) which are fed into the cross-entropy (CE) loss [6]:

$$\mathcal{L}_{\text{CE}} = -\frac{1}{N}\sum_{i=1}^{N} y_i \log(p_i) + (1 - y_i)\log(1 - p_i) \tag{2}$$

where N is the total number of predictions, y_i is the true class label (i.e. 0 or 1 for binary classification) of instance i and p_i is the predicted probability for the observation of instance i. Here, we note that the CE loss takes into account the probability of each classification, whereas the F1-score only considers the final classification label. Therefore, it is an important metric to consider when one is interested in robust predictions, as is needed for an effective fashion industry solution for reducing the number of returns.

In order to train the GNN discussed in the following section, an extra step is included into this methodology whereby the purchase events are only trained on if the product variant involved has an average return rate of higher than 80% or lower than 20%, in order to provide more robust positive and negative examples of return instances to the GNN. The reason for this is to investigate and avoid issues involving oversmoothing in the representations learnt by the GNN; however, all results are quoted on the entire test set with no filtering. The result of this is a dataset with 200,000 purchase events and an average vertex degree for the real nodes of 5 for product variant nodes and 2 for customer nodes.

4 Experiment Results

Table 1 displays the precision, recall and F1-scores each model evaluated on the full test dataset (1.5 M purchase events). The final hyperparameter values are chosen based on a validation set, randomly and uniformly constructed from 10% of the training data and are listed as: logistic regression ($C = 5.0$, tol. $= 10^{-4}$), MLP (# of layers $= 2$, hidden dim. $= 128$), random forest (# of estimators $= 100$, max. depth $= 6$, min. samples split $= 2$, min. samples leaf $= 1$, max. leaf nodes $= 10$), XGBoost [2] (booster $=$ gbtree, max. depth $= 4$, $\eta = 0.1$, $\gamma = 1$, min. child weight $= 1$, $\lambda = 2$, objective $=$ Binary Logistic, early stopping rounds $= 5$), GNN (1 GraphSAGE [8]

Table 1 Results for models considered in this section evaluated on the full test data

Model	Test scores			
	Precision	Recall	F1-score	CE loss $\mathcal{L}_{CE}$
Logistic regression	0.723	0.726	0.725	0.602
Random forest	0.788	0.712	0.748	0.630
MLP	**0.870**	0.656	0.748	0.582
XGBoost	0.805	**0.745**	0.774	0.561
GNN	0.855	0.737	**0.792**	**0.489**

Bold indicates the top performing model, where F1-score is maximised and the loss score is minimised

layer with dim. = 16, all aggregations = max. pool, dropout = 0.2, normalise = True).[4] For the MLP (16,641 trainable parameters) and GNN (49,665 trainable parameters) models, an Adam optimizer is used with a learning rate of 0.01.

The results in Table 1 show a superior performance for a GNN-based approach trained on high and low returning examples (described in Sect. 3) for F1-score and CE Loss metrics, indicating that a graph-based approach yields a better performing and more robust classification model. For reference, when comparing the same GNN to one trained on all available data, an F1-score of 0.783 was found, suggesting the GNN's performance may suffer from oversmoothing when being trained on less discrete positive and negative examples. Furthermore, as mentioned in Sect. 3, the classifier head attached to the GNN is the same MLP model also present in Table 1, therefore supporting the expectation that the graph embeddings from the GNN are able to encode useful information from the data. Table 1 also suggests that the GNN's predictions are more robust, based on a lower final CE loss (Equation (2)) combined with a higher F1-score evaluated on the test set.

Table 2 displays the F1-scores evaluated on the test set for individual country markets. In all country instances, the GNN-based approach obtains a superior F1-score to all other models considered. When comparing the results in these tables with the correlations discussed in Fig. 2, one can observe that those countries with higher correlations to a particular return label (1 or 0) are among the top performing F1-scores in Table 2.

Single market results are of particular interest to the wider e-commerce fashion industry in order to understand how to deliver the best service to customers and products across different individual markets. The ability to obtain results such as these are an important and unique feature in the novel ASOS GraphReturns dataset as it facilitates a level of understanding into how an ML model is performing across different areas and identify it is weaknesses. Note that a similar analysis can be done for different brands or product types.

---

[4] Any parameters not listed here are left at their default values provided by the packages sklearn [12] (Logistic Regression & Random Forest), xgboost [2] (XGBoost), PyTorch [11] (MLP). and PyG [5] (GNN).

Table 2 Summary of F1-scores and CE losses ($\mathcal{L}_{CE}$) evaluated on the test data for each individual country market

Model	Country A		Country B		Country C		Country D	
Trained on all markets								
	F1-score	$\mathcal{L}_{CE}$	F1-score	$\mathcal{L}_{CE}$	F1-score	$\mathcal{L}_{CE}$	F1-score	$\mathcal{L}_{CE}$
Logistic regression	0.635	0.611	0.776	0.606	0.658	0.611	0.593	0.608
Random forest	0.655	0.633	0.785	0.633	0.672	0.635	0.606	0.633
MLP	0.680	0.527	0.792	0.527	0.691	0.528	0.626	0.518
XGBoost	0.731	0.556	0.806	0.567	0.717	0.567	0.664	0.561
GNN	**0.757**	**0.436**	**0.821**	**0.487**	**0.744**	**0.485**	**0.732**	**0.494**
Model	Country E		Country F		Country G		Country H	
Trained on all markets								
	F1-score	$\mathcal{L}_{CE}$	F1-score	$\mathcal{L}_{CE}$	F1-score	$\mathcal{L}_{CE}$	F1-score	$\mathcal{L}_{CE}$
Logistic regression	0.812	0.591	0.729	0.618	0.673	0.605	0.671	0.610
Random forest	0.817	0.624	0.745	0.638	0.717	0.630	0.683	0.636
MLP	0.819	0.514	0.754	0.542	0.727	0.520	0.696	0.528
XGBoost	0.827	0.561	0.772	0.573	0.751	0.561	0.728	0.563
GNN	**0.842**	**0.487**	**0.801**	**0.500**	**0.774**	**0.489**	**0.744**	**0.505**

In these results, we use a GNN model with 1 SAGEGraph layer (dim. = 16) trained with all extra nodes considered from Sect. 3. Bold indicates the top performing model, where F1-score is maximised and the loss score is minimised

5 Conclusion

In this work, we have presented a novel dataset to inspire new directions in fashion retail research. This dataset is particularly suited to graph representation learning techniques and exhibits a naturally rich geometrical structure.

The baseline models which have been presented here to provide an early benchmark trained on the presented data support the claim that a GNN-based approach achieves a higher yield over the metrics considered. The best performing model is a GNN model described in Sects. 3 and 4 which obtained a final F1-score of 0.792 and a test CE loss score of 0.489 when evaluated on the test set. These results are an improvement from the next best performing model (2% higher F1-score and 6% lower CE loss) indicating the potential for graph-based methods on this naturally graph structured data. Of particular interest for e-commerce companies is the level of confidence when making a prediction which will affect the likelihood of a customer being notified by the prediction. Therefore, the final test CE loss value for the GNN being lower than other models implies that overall the GNN is likely more confident about its classifications than the other non-graph-based approaches. In order to reinforce this point, a future analysis of these predictions could include the investigation of calibrated probabilities as in [7].

As discussed, our primary goal is to provide a novel dataset to facilitate future research studies in fashion retail. This data is presented with labelled purchase links between customers and product variants which can be used in a supervised learning setting (as in Sect. 4). However, due to the graph structure of this data, it is possible to also use this data in the unsupervised setting with a wider range of transformer-based models. Finally, we wish to highlight the potential application of this dataset to advancements in recommendation systems. With the definite labels provided in this dataset which label a return, a future research direction would be investigating the universality of the GNN embeddings and how these translate into new recommendation systems for sustainable fashion.

References

1. Breiman L (2001) Random forests. Mach Learn 45:5–32. https://doi.org/10.1023/A:1010933404324
2. Chen T, Guestrin C (2016) Xgboost: A scalable tree boosting system. CoRR abs/1603.02754:785-794, arxiv.org/abs/1603.02754
3. Derrow-Pinion A, She J, Wong D, Lange O, Hester T, Perez L, Nunkesser M, Lee S, Guo X, Wiltshire B, Battaglia PW, Gupta V, Li A, Xu Z, Sanchez-Gonzalez A, Li Y, Velickovic P (2021) Eta prediction with graph neural networks in google maps. In: Proceedings of the 30th ACM international conference on information & knowledge management, CIKM'21. Association for Computing Machinery, New York, NY, USA, pp 3767–3776. https://doi.org/10.1145/3459637.3481916
4. Eksombatchai C, Jindal P, Liu JZ, Liu Y, Sharma R, Sugnet C, Ulrich M, Leskovec J (2018) Pixie: a system for recommending 3+ billion items to 200+ million users in real-time. In: Proceedings of the 2018 world wide web conference, international world wide web conferences steering committee, CHE, WWW'18. Republic and Canton of Geneva, pp 1775–1784. https://doi.org/10.1145/3178876.3186183
5. Fey M, Lenssen JE (2019) Fast graph representation learning with PyTorch Geometric. In: ICLR workshop on representation learning on graphs and manifolds
6. Good IJ (1952) Rational decisions. J Roy Stat Soc Ser B (Methodol) 14(1):107–114. http://www.jstor.org/stable/2984087
7. Guo C, Pleiss G, Sun Y, Weinberger KQ (2017) On calibration of modern neural networks. In: Proceedings of the 34th international conference on machine learning, ICML17, vol 70. JMLR.org, pp 1321–1330
8. Hamilton W, Ying Z, Leskovec J (2017) Inductive representation learning on large graphs. In: Guyon I, Luxburg UV, Bengio S, Wallach H, Fergus R, Vishwanathan S, Garnett R (eds) Advances in neural information processing systems, vol 30. Curran Associates, Inc. https://proceedings.neurips.cc/paper/2017/file/5dd9db5e033da9c6fb5ba83c7a7ebea9-Paper.pdf
9. He R, McAuley J (2016) Ups and downs: modeling the visual evolution of fashion trends with one-class collaborative filtering. In: Proceedings of the 25th international conference on world wide web, international world wide web conferences steering committee. https://doi.org/10.1145/2872427.2883037

10. Jumper J, Evans R, Pritzel A, Green T, Figurnov M, Ronneberger O, Tunyasuvunakool K, Bates R, Žídek A, Potapenko A et al (2021) Highly accurate protein structure prediction with alphafold. Nature 596(7873):583–589

11. Paszke A, Gross S, Massa F, Lerer A, Bradbury J, Chanan G, Killeen T, Lin Z, Gimelshein N, Antiga L, Desmaison A, Kopf A, Yang E, DeVito Z, Raison M, Tejani A, Chilamkurthy S, Steiner B, Fang L, Bai J, Chintala S (2019) Pytorch: an imperative style, high-performance deep learning library. In: Wallach H, Larochelle H, Beygelzimer A, d' Alché-Buc F, Fox E, Garnett R (eds) Advances in neural information processing systems, vol 32. Curran Associates, Inc., pp 8024–8035, http://papers.neurips.cc/paper/9015-pytorch-an-imperative-style-high-performance-deep-learning-library.pdf

12. Pedregosa F, Varoquaux G, Gramfort A, Michel V, Thirion B, Grisel O, Blondel M, Prettenhofer P, Weiss R, Dubourg V, Vanderplas J, Passos A, Cournapeau D, Brucher M, Perrot M, Duchesnay E (2011) Scikit-learn: machine learning in Python. J Mach Learn Res 12:2825–2830

13. Sanchez-Gonzalez A, Godwin J, Pfaff T, Ying R, Leskovec J, Battaglia PW (2020) Learning to simulate complex physics with graph networks. In: Proceedings of the 37th international conference on machine learning, ICML, ICML'20. JMLR.org

14. Stokes JM, Yang K, Swanson K, Jin W, Cubillos-Ruiz A, Donghia NM, MacNair CR, French S, Carfrae LA, Bloom-Ackermann Z et al (2020) A deep learning approach to antibiotic discovery. Cell 180(4):688–702

15. Wu S, Sun F, Zhang W, Xie X, Cui B (2022) Graph neural networks in recommender systems: a survey. ACM Comput Surv

End-to-End Image-Based Fashion Recommendation

Shereen Elsayed, Lukas Brinkmeyer, and Lars Schmidt-Thieme

Abstract In fashion-based recommendation settings, incorporating the item image features is considered a crucial factor, and it has shown significant improvements to many traditional models, including but not limited to matrix factorization, auto-encoders, and nearest neighbor models. While there are numerous image-based recommender approaches that utilize dedicated deep neural networks, comparisons to attribute-aware models are often disregarded despite their ability to be easily extended to leverage items' image features. In this paper, we propose a simple yet effective attribute-aware model that incorporates image features for better item representation learning in item recommendation tasks. The proposed model utilizes items' image features extracted by a calibrated ResNet50 component. We present an ablation study to compare incorporating the image features using three different techniques into the recommender system component that can seamlessly leverage any available items' attributes. Experiments on two image-based real-world recommender systems datasets show that the proposed model significantly outperforms all state-of-the-art image-based models.

1 Introduction

In recent years, recommender systems became one of the essential areas in the machine learning field. In our daily life, recommender systems affect us in one way or another, as they exist in almost all the current websites such as social media, shopping websites, and entertainment websites. Buying clothes online became a widespread trend nowadays; websites such as Zalando and Amazon are getting bigger every day.

S. Elsayed (✉) · L. Brinkmeyer · L. Schmidt-Thieme
Information Systems and Machine Learning Lab, University of Hildesheim, Hildesheim, Germany
e-mail: elsayed@ismll.uni-hildesheim.de

L. Brinkmeyer
e-mail: brinkmeyer@ismll.uni-hildesheim.de

L. Schmidt-Thieme
e-mail: schmidt-thieme@ismll.uni-hildesheim.de

H. J. Corona Pampín and R. Shirvany (eds.), *Recommender Systems in Fashion and Retail*, Lecture Notes in Electrical Engineering 981, https://doi.org/10.1007/978-3-031-22192-7_7

In this regard, item images play a crucial role, no one wants to buy a new shirt without seeing how it looks like. Many users also can have the same taste when they select what they want to buy; for example, one loves dark colors most of the time, others like sportswear, and so on. In fashion recommendation adding the items images to the model has proven to show a significant lift in the recommendation performance when the model is trained not only on the relational data describing the interaction between users and items but also on how it looks. It can massively help the model recommend other items that look similar or compatible. Some recent works tackled this area in the last years, particularly adding item images into the model with different techniques.

Following this research direction, we propose a hybrid attribute-aware model that relies on adding the items' image features into the recommendation model.

The contributions through this works can be specified as follows;

- We propose a simple image-aware model for item recommendation that can leverage items' image features extracted by a fine-tuned ResNet50 component [3].
- We conducted extensive experiments on two benchmark datasets, and results show that the proposed model was able to outperform more complex state-of-the-art methods.
- We conduct an ablation study to analyze and compare three different approaches to include the items images features extracted using the ReNet50 into the recommender model.

2   Related Work

Many image-based recommender systems were proposed in the last few years. They are becoming more popular, and their applications are getting wider, especially in fashion e-commerce websites. Many of the proposed models in the literature relied on using **pre-trained networks** for items images features extraction. In 2016 an essential state-of-the-art model was proposed, VBPR [4]. It uses the BPR ranking model [13] for prediction and takes the visual item features into account. They use a pre-trained CNN network to extract the item features; these features pass through a fully connected layer to obtain the latent embedding. Another model which uses more than one type of external information is JRL [15]. It incorporates three different information sources (user reviews, item images, and ratings) using a pre-trained CaffeNet and PV-DBOW model [8]. While in 2017, Qiang et al. [9] proposed the DeepStyle model, where their primary assumption is that the item representation consists of style and item category representations extracting the item image features via a CaffeNet model. To get the style features, they subtract latent factors representing the category. Additionally, a region-guided approach (SAERS) [5] introduced the items' visual features using AlexNet to get general features and utilizing a ResNet50 architecture for extracting semantic features representing the region of interest in the item image. Before semantic features are added to the global features, an attention mechanism using the users' preferences is applied. The final item embedding becomes the semantic features combined with the global features.

The image networks used for item image feature extraction can also be **trained end-to-end** with the recommender model; the most popular model applied this technique is the DVBPR [6] powerful model proposed in 2017 that incorporates visual item features. It does two tasks; the first task is training the BPR recommender model jointly with a CNN structure to extract the item's image pixel-level features. The second task uses Generative Adversarial Networks (GANs) to generate new item images based on user preferences.

Attribute-aware recommender system models are a family of hybrid models that can incorporate external user and item attributes. Theoretically, some of these models can be extendable to image-based settings by carefully converting the raw image features into real-valued latent features, which can be used as the item's attributes. Recently, (CoNCARS) [1] model was proposed that takes the user and item one-hot- encoded vectors as well as the user/item timestamps. The mode utilizes a convolution neural network (CNN) on top of the interaction matrix to generate the latent embeddings. Parallel work by Rashed et al. proposed an attribute-aware model (GraphRec) [11] that appends all the users' and items' attributes on-hot-encoded vectors. It extracts the embeddings through neural network layers that can capture the non-linearity in the user-item relation.

In the literature, using attribute-aware models has been mainly set aside for image-based items ranking problems. Hence, in this paper, we propose a simple image-aware model that utilizes the latent image features as item attributes. The experimental results show that the proposed model outperforms current complex image-based state-of-the-art models.

3 Methodology

3.1 Problem Definition

In image-based item recommendation tasks, there exist a set of M users $\mathcal{U} :=$ $\{u_1, \ldots, u_M\}$, a set of N items $\mathcal{I} := \{i_1, \ldots, i_N\}$ with their images $X_i \in \mathbb{R}^{N \times (L \times H \times C)}$ of dimensions $L \times H$ and C channels, and a sparse binary interaction matrix $R \in \mathbb{R}^{M \times N}$ that indicate user's implicit preferences on items based on historical interactions.

The recommendation task's primary goal is to generate a ranked personalized short-list of items to users by estimating the missing likelihood scores in R while considering the visual information that exists in the item images.

3.2 Proposed Model

The proposed model consists of an image features extraction component and an attribute-aware recommender system component that are jointly optimized.

3.2.1 Recommender System Component

Inspired by the GraphRec model [11], the recommender system component utilizes the user's one-hot encoded input vector and concatenates the external features of the items directly to the items' one-hot input vectors. These vectors are then fed to their independent embedding functions $\psi_u : \mathbb{R}^M \rightarrow \mathbb{R}^K$ and $\psi_i : \mathbb{R}^{(N+F)} \rightarrow \mathbb{R}^K$ as follows:

$$z_u = \psi_u(v_u) = v_u W^{\psi_u} + b^{\psi_u} \tag{1}$$

$$z_i = \psi_i(v_i) = \text{concat}(v_i, \phi(x_i)) W^{\psi_i} + b^{\psi_i} \tag{2}$$

where W^{ψ_u}, W^{ψ_i} are the weight matrices of the embedding functions, and b^{ψ_u}, b^{ψ_i} are the bias vectors. v_u, v_i represents the user and item one-hot encoded vectors. Additionally, $\phi(x_i)$ represents the features extraction component that embeds an item's raw image x_i to a latent feature vector of size F.

After obtaining the user and item embeddings, the final score is calculated using the dot-product of the two embedding vectors, $\hat{y}_{ui} = z_u \cdot z_i$ to give the final output score representing how much this user u will tend to like this item i. The final score is computed via a sigmoid function; $\sigma(\hat{y}_{ui}) = 1/1 + e^{(\hat{y}_{ui})}$, to limit the score value from $0 \rightarrow 1$. The model target is defined as y_{ui} which is for implicit feedback either 0 or 1;

$$y_{ui} = \begin{cases} 1, & \text{observed item;} \\ 0, & \text{otherwise} \end{cases} \tag{3}$$

Given the users-items positive and negative interactions D_s^+, D_s^-, and the output score $\hat{y}_{ui}$ of the model and the original target y_{ui}, we use negative sampling for generating unobserved instances and optimize the negative log-likelihood objective function $\ell(\hat{y}; D_s)$ using ADAM optimizer, which can be defined as;

$$- \sum_{(u,i) \in D_s^+ \cup D_s^-} y_{ui} \log(\hat{y}_{ui}) + (1 - y_{ui})(1 - \log(\hat{y}_{ui})) \tag{4}$$

3.2.2 Extraction of Image Features

To extract the latent item's image features, we propose using the ResNet50 component for combining the raw image features. To refine the image features further and get better representation, we can jointly train the whole image network simultaneously with the recommender model. However, it will require a considerable amount of memory and computational power to load and update these parameters. In this case, ResNet50 consists of 176 layers with around 23 million parameters. To mitigate this problem, we propose ImgRec End-to-End (ImgRec-EtE), where we utilize a ResNet50 [3] pre-trained on ImageNet dataset [2] and jointly train part of the image network to be updated with the recommender model, and at the same time, benefit from starting with the pre-trained weights. As shown in Fig. 1, we selected the last 50 layers to be updated and fix the first 126 layers. Furthermore, we added an additional separate, fully connected layer to fine-tune the image features extracted by the ResNet50. This layer will be trained simultaneously with the recommender model. Moreover, this additional layer makes the image features more compact and decreases its dimensionality further to match the user latent embedding. Thus the features extraction function $\phi(x_i)$ for ImgRec-EtE can be defined as follows;

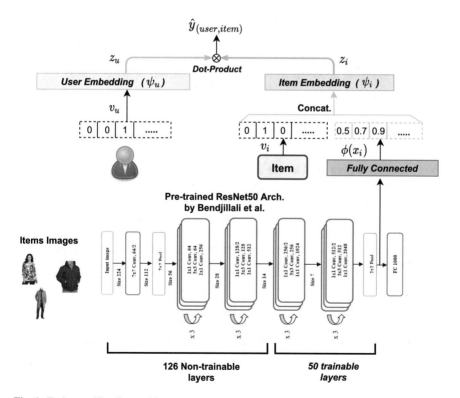

Fig. 1 End-to-end ImgRec architecture

$$\phi(x_i) := \text{ReLU}(\text{ResNet50}(x_i)W^{\phi} + b^{\phi}) \tag{5}$$

3.3   Training Strategy

To increase the speed of the training process, we used a two-stage training protocol. Firstly, we train the model by fixing the image network pre-trained parameters and updating the recommender model parameters until convergence. After obtaining the best model performance in the first stage, we jointly learn and fine-tune the last 50 layers of the image network further with the recommender model. This methodology allowed us to fine-tune the model after reaching the best performance given the pre-trained image network weights, also it saves time and computational power while achieve superior prediction performance compared to using only the fixed pre-trained parameters of the image network.

4   Experiments

Through the experimental section, we aim to answer the following research questions:

RQ1 How does the proposed model fair against state-of-the-art image-based models?

RQ2 What is the best method for adding images features to the model?

4.1   Datasets

We chose two widely used image-based recommendation datasets *Amazon fashion* and *Amazon men*, introduced by McAuley et al. [10]. *Amazon fashion* was collected from six different categories, "men/women's tops, bottoms, and shoes," while *Amazon men* contains all subcategories (gloves, scarves, sunglasses, etc.). The number we mention of users-items of the Fashion dataset is different from the ones stated in the main paper. However, we contacted the authors[1] and the numbers in Table 1 were found to be the correct ones.

Table 1 Datasets statistics

Dataset	Users	Items	Categories	Interactions
Amazon fashion	45,184	166,270	6	267,635
Amazon men	34,244	110,636	50	186,382

---

[1] https://github.com/kang205/DVBPR/issues/6

4.2 Evaluation Protocol

To evaluate our proposed models, the data is split into training, validation, and test sets using the leave-one-out evaluation protocol as in [4–6]. However, we used two different numbers of negative samples in the evaluation protocol for the direct comparison against the published results of the state-of-the-art models DVBPR [6], and SAERS [5] because both models originally used a different number of samples, and the source code of SAERS is not available.

For the direct comparison against DVBPR, we sample 100 negative samples ($\mathcal{I}^t$) and one positive item i for each user. On the other hand, for direct comparison against our second baseline SAERS [5] we sample 500 negative items ($\mathcal{I}^t$) and one positive item i. To ensure our results' consistency, we report the mean of each experiment's five different trials.

For evaluation, we report the **Area Under the Curve (AUC)** as it is the primary metric in all of our baselines' papers, further more it is persistent to the number of negative samples used [7]:

$$\text{AUC} = \frac{1}{|\mathcal{U}|} \sum_{u \in \mathcal{U}} \frac{1}{|\mathcal{I}^t|} \sum_{i,j \in \mathcal{I}^t} (\hat{y}_{ui} > \hat{y}_{uj}) \tag{6}$$

4.3 Baselines

We compared our proposed methods to the published results of the state-of-the-art image-based models DVBPR and SAERS. We also compared our results against a set of well-known item recommendation models that were used in [5, 6].

- **PopRank**: A naive popularity-based ranking model.
- **WARP** [14]: A matrix factorization model that uses Weighted Approximate-Rank Pairwise (WARP) loss.
- **BPR-MF** [13]: A matrix factorization model that uses the BPR loss to get the ranking of the items.
- **VisRank**: A content-based model that utilizes the similarity between CNN features of the items bought by the user.
- **Factorization Machines (FM)** [12]: A generic method that combines the benefits of both SVM and factorization techniques using pair-wise BPR loss.
- **VBPR** [4]: A model that utilizes items; visual features, using pre-trained CaffeNet and a BPR ranking model.
- **DeepStyle** [9]: A model that uses the BPR framework and incorporates style features extracted by subtracting category information from CaffeNet visual features of the items.
- **JRL** [15]: A model that incorporates three different types of item attributes. In this case, we considered only the visual features for comparison purposes.

- **DVBPR** [6]: State-of-the-art image-based model that adds the visual features extracted from a dedicated CNN network trained along with a BPR recommender model.
- **SAERS** [5]: State-of-the-art image-based model that utilizes the region of interests in the visual images of the items while also considering the global extracted features from a dedicated CNN to get the final items representations.

4.4 Comparative Study Against State-of-the-Art Image-Based Models (RQ1)

Since the DVBPR baseline conducted their results on Amazon men and Amazon fashion datasets, we compared our results directly to both datasets' published results. On the other hand, the SAERS model used only the Amazon fashion dataset, so we only report the results for this dataset using 500 negative samples per user. Table 2 illustrates the ImgRec-EtE obtained results against VBPR and DVBPR results. The proposed model ImgRec-EtE represents the best performance on both men and fashion datasets. It shows a 2.5% improvement over the fashion dataset DVBPR reported performance and a 4.8% improvement for the men dataset. The results show consistent AUC values regardless of the number of negative samples, as per the recent study by Krichene et al. [7]. Table 3 demonstrates the comparison against the Deep-Style, JRL, and SAERS models. The proposed model ImgRec-EtE represents the best performance on the fashion dataset. Despite its simplicity, the model has achieved an AUC of 0.825, which shows an improvement over the complex state-of-the-art SAERS model with 0.9%.

Table 2 Comparison of AUC scores with 100 negative samples per user, the bold results represent the best performing model and we underline the second best result

Datasets	Interactions			
	PopRank	WARP	BPR-MF	FM
Amazon fashion	0.5849	0.6065	0.6278	0.7093
Amazon men	0.6060	0.6081	0.6450	0.6654
Datasets	Interactions + Image features			
	VisRank	VBPR	DVBPR	ImgRec-EtE
Amazon fashion	0.6839	0.7479	<u>0.7964</u>	**0.8250**
Amazon men	0.6589	0.7089	<u>0.7410</u>	**0.7899**

Table 3 Comparison of AUC scores with 500 negative samples per user

Datasets	Interactions		Interactions + Image features				
	PopRank	BPR-MF	VBPR	DeepStyle	JRL	SAERS	ImgRec-EtE
Amazon fashion	0.5910	0.6300	0.7710	0.7600	0.7710	<u>0.8161</u>	**0.8250**

4.5 Ablation Study (RQ2)

Besides obtaining the items features in an end-to-end fashion, it is worth mentioning that we tried other methods to incorporate the images' features. Firstly in (ImgRec-Dir), we directly concatenate the image features extracted using the output of the next to last fully connected layer of a pre-trained ResNet50 to the one-hot encoded vector representing the item. On the other hand (ImgRec-FT) passes the features extracted using the pre-trained network to a fine-tuning layer that is trained with the recommender model and obtain better item representation. Subsequently, the item's image latent features are concatenated to the item one-hot encoded vector to form one input vector representing the item. As shown in Table 4 The images' features had a varying effect depending on how they were added to the model; ImgRec-Dir achieved an AUC of 0.77 on the Amazon fashion dataset and 0.736 on the Amazon men dataset. While looking into ImgRec-FT performance after adding the fine-tuning layer, we can see an improvement of 3.2% on the Amazon fashion dataset and 1.6% on the Amazon men dataset performances, which shows high competitiveness against the state-or-the-art models while having a much lower computational complexity. Finally, ImgRec-EtE, which jointly trains part of the ResNet50 simultaneously with the model, positively impacted the results with further improvement of 1.6% over the ImgRec-FT performance on both datasets.

Table 4 Comparison of AUC scores with 100 negative samples per user, between the three ways of incorporating the image features

Datasets	ImgRec-Dir	ImgRec-FT	ImgRec-EtE
Amazon fashion	0.7770	0.8090	**0.8250**
Amazon men	0.7363	0.7735	**0.7899**

4.6 Hyperparameters

We ran our experiments using GPU RTX 2070 Super and CPU Xeon Gold 6230 with RAM 256 GB. We used user and item embedding sizes of 10 and 20 with *Linear* activation function for both datasets. We applied grid search on the learning rate between [0.00005 and 0.0003] and the L2-regularization lambda between [0.000001 and 0.2]. The best parameters are 0.0001 and 0.1 for ImgRec-Dir and ImgRec-FT. While in ImgRec-EtE case, the best L2-regularization lambda is 0.000001 for phase1 (fixed-weights) and 0.00005 for phase 2 (joint training). The features fine-tuning layer, the best-selected embedding size is 150 with ReLU activation function. ImgRec codes and datasets are available at https://github.com/Shereen-Elsayed/ImgRec.

5 Conclusion

In this work, we propose an image-based attribute-aware model for items' personalized ranking with jointly training a ResNet50 component simultaneously with the model for incorporating image features into the recommender model. Adding the image features showed significant improvement in the model's performance. ImgRec-EtE shows superior performance to all image-based recommendation approaches. Furthermore, we conducted an ablation study to compare different approaches of adding the features to the model; direct features concatenation, adding a fine-tuning fully connected layer, and jointly training part of the image network.

Acknowledgements This work is co-funded by the industry Project "IIP-Ecosphere: Next Level Ecosphere for Intelligent Industrial Production".

References

1. Costa FSd, Dolog P (2019) Collective embedding for neural context-aware recommender systems. In: Proceedings of the 13th ACM conference on recommender systems, pp 201–209
2. Deng J, Dong W, Socher R, Li LJ, Li K, Fei-Fei L (2009) ImageNet: a large-scale hierarchical image database. In: CVPR09
3. He K, Zhang X, Ren S, Sun J (2016) Deep residual learning for image recognition. In: Proceedings of the IEEE conference on computer vision and pattern recognition, pp 770–778
4. He R, McAuley J (2015) VBPR: visual Bayesian personalized ranking from implicit feedback. arXiv:1510.01784
5. Hou M, Wu L, Chen E, Li Z, Zheng VW, Liu Q (2019) Explainable fashion recommendation: a semantic attribute region guided approach. arXiv:1905.12862
6. Kang WC, Fang C, Wang Z, McAuley J (2017) Visually-aware fashion recommendation and design with generative image models. In: 2017 IEEE international conference on data mining (ICDM). IEEE, pp 207–216
7. Krichene W, Rendle S (2020) On sampled metrics for item recommendation. In: Proceedings of the 26th ACM SIGKDD international conference on knowledge discovery & data mining, pp 1748–1757

8. Le Q, Mikolov T (2014) Distributed representations of sentences and documents. In: International conference on machine learning, pp 1188–1196
9. Liu Q, Wu S, Wang L (2017) Deepstyle: Learning user preferences for visual recommendation. In: Proceedings of the 40th international ACM SIGIR conference on research and development in information retrieval, pp 841–844
10. McAuley J, Targett C, Shi Q, Van Den Hengel A (2015) Image-based recommendations on styles and substitutes. In: Proceedings of the 38th international ACM SIGIR conference on research and development in information retrieval, pp 43–52
11. Rashed A, Grabocka J, Schmidt-Thieme L (2019) Attribute-aware non-linear co-embeddings of graph features. In: Proceedings of the 13th ACM conference on recommender systems, pp 314–321
12. Rendle S (2010) Factorization machines. ICDM
13. Rendle S, Freudenthaler C, Gantner Z, Schmidt-Thieme L (2012) BPR: Bayesian personalized ranking from implicit feedback. arXiv:1205.2618
14. Weston J, Bengio S, Usunier N (2011) WSABIE: scaling up to large vocabulary image annotation
15. Zhang Y, Ai Q, Chen X, Croft WB (2017) Joint representation learning for top-n recommendation with heterogeneous information sources. In: Proceedings of the 2017 ACM on conference on information and knowledge management, pp 1449–1458

Printed in the United States
by Baker & Taylor Publisher Services